unications Series
nes Edited by
H. Sterling

communications Infrastructure

ism and Telecommunications

o Know: Media, Democracy,

The LEA Telecomm
A Series of Volu
Christopher

AMERICAN REGULATORY FEDERALISM AND TELECOMMUNICATIONS INFRASTRUCTURE

Edited by

Paul Teske
SUNY Stony Brook

LEA LAWRENCE ERLBAUM ASSOCIATES, PUBLISHERS
1995 Hillsdale, New Jersey Hove, UK

Lawrence Erlbaum Associates, Inc., Publishers
365 Broadway
Hillsdale, New Jersey 07642

Library of Congress Cataloging-in-Publication Data

American regulatory federalism and telecommunications infrastructure /
   edited by Paul Teske.
        p.   cm.
   Includes bibliographical references and indexes.
   ISBN 0-8058-1615-1
      1. Telecommunication policy—United States.   2. Telecommunication—
Law and legislation—United States.   3. Telephone—Law and
legislation—United States.   I. Teske, Paul.
HE7781.A58 1994
384'.068—dc20   94-28042
                                                              CIP

Books published by Lawrence Erlbaum Associates are printed on acid-free
paper, and their bindings are chosen for strength and durability.

Printed in the United States of America
10 9 8 7 6 5 4 3 2 1

CONTENTS

PART IV: FEDERALISM AND THE FUTURE

ACKNOWLEDGMENTS

Primarily, I would like to thank Eli Noam, Barry Cole, and Doug Conn of the Columbia University Institute for Tele-Information for agreeing to sponsor a conference on state regulation, with the aim of preparing an edited book. Most of the authors of chapters in this book presented their initial work at that conference. I would also like to thank the Department of Political Science at SUNY Stony Brook for their support of this research. Richard Chard did an excellent job preparing the index for the book. On the publishing side, I much appreciated Christopher Sterling's early enthusiasm about the book for the Erlbaum series. The professional staff at Lawrence Erlbaum Associates, in particular Hollis Heimbouch and Debbie Ruel, have been very helpful in improving the final product.

OVERVIEW AND HISTORY

1

INTRODUCTION AND OVERVIEW

Paul Teske
State University of New York
at Stony Brook

For more than a century, scholars, judges, and politicians have sought an appropriate mix of national and state jurisdiction in the regulation of core "infrastructure" industries, such as railroads, electricity, telecommunications, insurance, and banking. Although the U.S. Constitution granted the federal government clear authority over *interstate* commerce, Congress took its first specific legislative initiative with the 1887 Interstate Commerce Act, to establish federal government jurisdiction over interstate commerce regulation, particularly for railroads. In that act and numerous other subsequent federal laws and regulation, control over *intrastate* regulation theoretically has been reserved to the states, unless such state regulation violates compelling national interests.

In reality, the mix of such interstate and intrastate powers has changed over time and varies greatly by industry. For example, insurance regulation is performed almost exclusively by the states, whereas railroad regulation is now performed almost exclusively by the federal government. Despite such variation, the overall U.S. trend of the past century is best described as sporadic movements that have favored centralized, federal regulatory power. However, significant elements of state regulation remain in most regulated industries and state regulation is certainly very important in telecommunications as of 1994.

Outside of the U.S. context, issues of localized versus centralized economic and regulatory control are also highly topical in the European Economic Community, as well as in the nations of Eastern Europe and the emerging independent republics formerly under the control of the USSR. In this book, chapter contributors analyze different aspects of U.S. state regulation and the current and prospective balance between central control and state regulatory independence in telecommunications. As several authors note, the United States is the only nation to continue important elements of subnational regulation. Are we lagging behind or setting an international trend?

POLICY ISSUES IN TWO-TIERED TELECOMMUNICATIONS REGULATION

More than a decade has passed since the United States decided to dismantle its largest corporation, AT&T, and fully embark on the experiment of telecommunications deregulation. In that time, other industrialized nations have made important changes in their telecommunications systems, but no other industrialized country has retained a two-tiered system of regulation. This is an appropriate time to assess the state-level institutions and implementation of U.S. policy. This book uses a positive political economy perspective to analyze enduring state–federal conflicts and to weigh the justifications and explanations for continuing state telecommunications regulation, or for changing its structure.

Since the actual divestiture of local operating companies from AT&T in 1984, states have varied greatly in the extent to which they have followed the central federal policy of changing rate structures, by reducing long-distance rates and increasing end-user charges, and opening markets to competitive entry (see Teske, 1990). Although some states have made major regulatory policy changes, many have taken only small steps. The states seem to offer a variety of policy models.

Although the Federal Communications Commission (FCC) regulates *interstate* activity, largely by long-distance carriers, and the states regulate *intrastate* communications (both local and long distance), the separation between the two does not neatly follow network boundaries or the emergence of newer services, leading to legal battles over jurisdictional authority. As Eli Noam and Henry Geller discuss later, both before and immediately after divestiture the FCC successfully preempted the authority of state public utility commissions (PUCs) in a number of rulings.[1] This trend was abruptly reversed in 1986, with the U.S. Supreme Court *Louisiana* ruling, in favor of the states on the issue of intrastate depreciation rates. Congressional action in 1993, however, preempting state regulation of cellular and wireless service rates may set the tone for further Congressional preemption of state authority, which is much more likely to withstand court scrutiny than FCC preemption.

Regulation of the telecommunications industry is further complicated by the continuing presence of D.C. District Court Judge Harold Greene, who presided over the AT&T divestiture and is still actively involved in monitoring related

---

[1]State telephone regulatory bodies are most often called public utility commissions (PUCs) or public service commissions (PSCs). In a few states these regulatory agencies have different names, such as Boards of Public Utilities and Corporation Commissions. In this book we use the most common term, PUCs, to describe all these state telecommunications regulatory bodies.

antitrust issues. While his and other legal decisions shape the boundaries of regulation, the truly important questions discussed in this book are not mainly legal concerns; they involve critical political and economic debates about cost burdens for different categories of consumers, market entry for different firms, economic growth and the information infrastructure, global competitiveness, and control over information. As the Clinton–Gore administration advances its goal of a National Information Infrastructure (NII) leading to an "information super-highway," the question of state regulation becomes even more important as state regulators retain many important decisions that shape the extent and timing of development of such an infrastructure.

In particular, the rate structure choices that state regulators face are very difficult to resolve because economic efficiency dictates changes that have significant political implications. The rate structure that evolved after World War II included a cross-subsidy to residential subscribers' local bills, paid for by long-distance users. Although the exact extent of this subsidy is highly controversial and is not easily resolved by economic theory, most economists call for increased end-user charges and reduced toll rates to price telecommunications efficiently (see, for example, Wenders, 1987). Such rate structure changes discourage large business users of telecommunications from bypassing the switched network (which tends to reduce the source of the subsidy), and will reduce their telecommunications bills substantially. However, these changes tend to increase total telephone bills for a majority of residential consumers, at least in the short run. The possibility of a majority of consumers facing higher bills obviously causes political resistance, and has made state regulators, to whom local increases are most easily traced, less likely to change rate structures than their federal counterparts. These issues are also difficult because of the competitive questions of who should be allowed to participate in which markets. Economies of scope may be best exploited by allowing the regulated local exchange companies (LECs) to participate in a wide range of new and enhanced information services, such as database access or video on demand. But the LECs still control the "bottleneck" local transport and switching functions that nearly all carriers must use to connect to ultimate users. Thus, the local companies may be able to choke off competitors and cross-subsidize services to their own advantage (which is why Judge Greene continues to monitor them closely for these antitrust issues). The LECs have argued since the early 1980s that they should be allowed into new markets, while less regulated competitors have begun to nibble into the margins of their formerly captive markets, particularly in the most profitable segments of the local market (see, for example, Teske & Gebosky, 1991).

CRITICISM OF STATE REGULATION

Part of the reason for examining the dual structure of U.S. regulation is to consider whether or not it promotes effective regulatory policy and industry development. Some argue that it is a superior system that has stood the test of time. Critics of the current two-tiered structure, on the other hand, suggest that 50 states regulating telecommunications leads to a fragmented national system, reduced service quality, incompatability, and unwieldy constraints on companies that are needed to carry the U.S. competitive banner abroad in international markets. Such critics also point out that before AT&T was dismantled it played a coordinating role around the country that no single institution, public or private, plays today. Ken Robinson, a top policy advisor to the heads of the Department of Commerce's National Telecommunications and Information Administration (NTIA) and the FCC, framed the question: "It is difficult to argue that state regulation is useful for advancing our national goals for these firms operating in global markets. . . . How likely is it that state regulation will remain relevant in this century?" (Teske, 1987, p. 6).

Such critics suspect that state regulatory agencies are not equipped with adequate expertise to evaluate these difficult policy choices and will inevitably make the most politically expedient, short-run choices. Robinson noted further: "It has certainly been true at the federal level for 20 years that state regulators have been viewed as poor relations. . . . Federal agencies have viewed them with suspicion and guarded distrust" (Teske, 1987, p. 6).

Noll (1986) argued that federal and state regulators are fundamentally in conflict. According to Noll, state regulators seek mainly to hold local rates down for political reasons, whereas the FCC would like to eliminate subsidies so that competitors can fight on a level playing field. Noll (1986) suggested: "In short, the FCC, to the extent it is legally and politically free to do so, should behave as if it were in serious conflict with state regulators over the ultimate aim of telecommunications policy" (p. 12). To some extent, the FCC has followed Noll's advice in recent years.

Noll (1989) went even further in his criticisms: "In the long run the telecommunications system might better serve society's objectives if, as in broadcasting, state regulation played no role at all; and as a practical matter the jurisdictional boundary between state and federal authorities is now quite blurry, so that debate about where it should be drawn is timely" (p. 18).

HISTORY OF TWO-TIERED REGULATION

Noll (1986) characterized state goals to keep residential prices low as leading to "the possibility of turning the BOCs into the railroads of the Twenty-First Century: sluggish, inefficient firms that are moribund because of the constraints imposed by regulation" (p. 5). Indeed, the current state of telecommunications regulation has striking analogies to previous U.S. regulation of transportation industries.

Over 100 years ago, important interests in the states debated how to regulate pricing and competition in railroads, with their high fixed costs (tracks) and low marginal costs.[2] Railroad tracks ran in and out of individual states, making it difficult for any one state to regulate effectively. Congress considered, but could not a pass for several years, bills that would establish federal regulation of interstate railroad transport. In 1886, the U.S. Supreme Court ruled, in *Wabash v. Illinois*, that states could not regulate railroads whose lines ran outside the state, even in the absence of federal regulation. This decision effectively forced Congressional action. In 1887, Congress passed a law establishing the Interstate Commerce Commission (ICC) to regulate railroads that ran interstate operations. State regulators continued to regulate intrastate commerce, although for most railroads this gradually became relatively unimportant as their commerce became 90% interstate, and in 1980 Congress preempted state regulation of railroads.

Another example of two-tiered regulation developed in trucking in the 1930s. Several states started regulating trucking before the 1920s at the strong urging of competing (regulated) railroad interests, but interstate trucking activity clearly was poised to grow, competing with railroads for long hauls. The railroad interests, combined with some larger trucking firms that saw federal regulation as likely to help them control entry, stabilize profits, and reduce the fragmentation of state regulation, pushed Congress to pass the Motor Carrier Act of 1935, which put interstate trucking under the regulatory control of the ICC. State regulators continued to regulate an intrastate trucking industry that is as large or larger than the interstate market (see Teske, Best, & Mintrom, 1994). Most state regulators continue to regulate intrastate trucking, despite the 1980 Congressional deregulation of interstate trucking, which did not put state regulators out of business as they were in railroad regulation.

Although the primacy of intrastate regulation declined relative to federal intervention in these transportation industries over time, the variation in the degree of state regulation played an important role in providing economic evidence that helped stimulate deregulation in the 1970s and 1980s. Econometric studies showed that regulated interstate trucking rates were 10%–15% higher than such rates in states with limited or no regulation. Similarly, some large states, particularly Texas and California, did not regulate intrastate airline competition as the federal Civil Aeronautics Board (CAB) did, and this allowed analysts to develop data showing the efficiency of these markets compared to the CAB's interstate airlines cartel.

State public utility commissions initiated telecommunications regulation in the early part of the 20th century, while federal policy was handled by the ICC. State telecommunications regulation spread rapidly (see Cohen, 1992). As Noam notes in this volume, the federal role was strengthened in the 1934 Communications Act with the establishment of the FCC, although this Act did not adequately clarify the jurisdictional issues between the federal government and the states.

The most contentious issue at that time, as it is today, was the rate structure. The 1931 U.S. Supreme Court case of *Smith v. Illinois Bell* argued that because

---

[2]See McCraw (1984) for an excellent history of early regulatory struggles with these questions.

toll calls required the existence of local networks, toll call prices should include a contribution toward paying these local costs. Today, most economists argue that such logic runs counter to economic efficiency. Still, this method was adopted and the subsidy from toll calls to residential service grew steadily after World War II.

This policy of subsidizing residential end users from toll call revenue aided the federal goal of universal service, and with the regulated natural monopoly protected from "cream-skimming" competition, it continued with few changes until competitors began to nibble at the edges of the Bell monopoly in the 1960s. With the development of microwave communications, and subsequent FCC decisions expanding long-distance competition in the 1970s, the previously stable environment began to unravel. Inflation and slower technological progress meant that local service costs no longer fell as rapidly as in the past. Technological innovation continued to drive toll costs down, however, so regulators, particularly Congress and state regulators, pushed to increase the subsidy from toll prices to local rates to retain the political advantage of falling local rates. With the court-mandated divestiture in 1982, however, this subsidy system faced severe constraints. The federal government and the states faced difficult decisions and a more complex relationship. The focus of this book is on how these choices have been handled and whether the two-tiered system has proved adaptive or not.

NORMATIVE JUSTIFICATIONS FOR STATE REGULATION

A political economy analysis of this dual system of regulation compares its costs and benefits to a single federal system of regulation that is the norm in almost all other nations. Such an analysis considers separately what social scientists call normative and positive evaluation. *Normative evaluation*, associated with economic analysis, considers what *should* be, which implies the need to establish criteria or values to evaluate that which is appropriate or inappropriate according to those criteria. Economic efficiency is the criterion agreed on by most economists. *Positive evaluation*, associated more with political science research traditions, tries to avoid value judgments and considers why we have the regulatory system we have, regardless of whether it is a good or bad system.

Starting with normative evaluation, at least in theory, state telecommunications regulation can provide real benefits to all parties. First, state regulation takes place "closer" to the citizens and businesses of a state. Thus, state regulators know more about their own economy and society, and can tailor their policies appropriately according to these local needs. Needs, preferences, and tastes for telecommunications services may vary across states, related to such issues as subsidizing poor and rural people onto networks, the desire to have competitive alternatives, and the trade-off of current prices versus future network services that is implicitly reflected in the arcane accounting details of depreciation rates.

As Andrew Varley, former chair of the Iowa State Utility Board, argued: "It is essential to allow state laboratories to experiment and see what works. . . . If you live in Washington, DC, you can reach with a local call roughly the same number of people as the entire population of the states of Iowa and North Dakota. . . . These people (FCC) should totally understand our problems before prescribing solutions" (Teske, 1987, p. 18).

In parallel, state regulation should also lead to more public participation in decisions and more responsiveness by politicians and their appointees to local interests. As Gormley argues herein, participation, responsiveness, and accountability are very important political goals. Traditionally, however, scholars have found business interests to be *more* dominant on the state level than on the federal level, because of less intense scrutiny of regulation by the media and consumer groups at the state level, and states' needs for economic development from mobile businesses (see Lindblom, 1977; Schattschneider, 1960). At least in the larger states, however, this situation has been changing. In the 1980s, consumer groups and the media realized that not all important regulatory activity takes place in Washington, DC, and earlier state procedural and institutional reforms reduced the likelihood of complete business dominance (Scholz, 1981).

Another important justification is that states can act as experimental laboratories, from which analysts, other states, and the federal government can learn. In the area of economic development, Osborne (1988), Fosler (1988), and Eisinger (1988) argued that states have provided important policy laboratories. This type of policy response fosters a kind of incrementalism in U.S. policy that, to paraphrase Lindblom (1977), can be called "the intelligence of federalism." According to Stanford Levin, former Illinois Commerce Commission regulator: "State diversity is helpful to learn more about the elements of uncertainty" (Teske, 1987, p. 7). Issue networks created through the National Association of Regulatory Utility Commissions (NARUC), the National Governors Association, and regional groups such as the Council of Great Lakes Governors, encourage such learning. According to Eliot Maxwell, executive director of external affairs for Pacific Telesis: "State regulation might converge to become more homogeneous in 10 years, but the differences today might help choose among policy approaches" (Teske, 1987, p. 8).

As noted earlier, state policy variation contributed greatly to the evidence in favor of federal transportation deregulation. In telecommunications, for example, a few states were ahead of the federal government on the increasingly important issue of network interconnection agreements, particularly New Jersey, New York, and Massachusetts. Clearly the FCC learned from those state policy experiments and developed its own policy with an eye on state successes and failures.

As a related subset to the policy experimentation issue, some states are attempting to use their regulatory innovations to stimulate economic development (see chaps. 3 and 4). Many analysts and firms argue that there is a critical link between economic development and state telecommunications regulatory policies.

Nebraska found this argument convincing enough to deregulate radically in the hope that it could enhance its status as the "1-800" number capital of the United States (see Teske, 1990). The benefits and costs of interstate economic development competition are the subject of widespread debate; some feel that competition promotes policy innovation that "expands the pie," whereas others suggest that it is usually a "zero-sum game" that creates a prisoner's dilemma for states, ultimately leading to taxpayer subsidies of private corporations. To the extent that the first view is accurate (see Eisinger, 1988), varying state regulatory models can illustrate the importance of the telecommunications infrastructure in stimulating economic growth (Wilson & Teske, 1990).

Although these theoretical benefits may justify ongoing state regulation, in theory such regulation also may increase costs to our society. Telecommunications networks require standards, so that all subscribers can communicate with each other, regardless of where they live and what local tastes are on various policy issues. We need an interconnected national network, rather than 51 (including DC, which has its own regulators) separate state networks, and this may become even more critical as newer and more complicated communications services become feasible on the information superhighway. Many observers feel that superior telecommunications is a vital infrastructure for international competition, and that state regulatory fragmentation, which has no counterpart in Japan or western Europe, hinders this development (Harris, 1988a, 1988b).

Integrated Systems Digital Network (ISDN) may be the best example of the need for standardization across local networks now regulated by different state commissions. As in the computer industry, however, component compatibility can come about in many ways—voluntary, enforced, or mixed models are possible. Although single, centrally planned technological standards are often touted as desirable, there may be trade-offs between enforced uniform standard setting and further technological innovation. A mixed model, with minimum standards, but room for innovation, as practiced on today's rapidly growing Internet, may be appropriate.

Noll outlined four options for standard setting disputes: "1) full federal control; 2) state control; 3) a consortium akin to the old Ozark plan . . . to develop consensus among all participants; or 4) minimum standards for elements important in the federal jurisdiction set by federal officials, with states deciding as they wish within those federally-imposed strictures" (quoted in Entman, 1988, p. 38). Noll predicted that this fourth option is most likely.

Issues of multiple jurisdictions and standards setting are not simply scholarly debates; comparative advantage is involved for many segments of the industry. According to Northern Telecom's Patrick Keenan, "Many of the potential conflicts between federal and state regulators are in fact driven by enhanced service providers who complain they do not want to work with 51 separate state jurisdictions" (quoted in Entman, 1988, p. 33). In another example: "Daniel Kelley of MCI said interexchange carriers, not just ESPs, have strong fears of too much variation across state jurisdictions" (Entman, 1988, p. 33).

Related to the standards issues, positive network externalities do not end at state boundaries, they can flow from one state to another, as with new information services. Therefore, a second justification for national regulation is the positive network externalities to all subscribers from expanded subscribership in *other* states. Any network becomes more valuable to all of its existing subscribers when more other subscribers can be reached; positive externalities were the implicit economic justification for the national goal of "universal service." If some states develop policies that lose subscribers, or that act to retard the expansion of new network services, customers in other states will be harmed by the smaller network size. Haring and Levitz (1989) cited these externalities as the major consideration opposing the advantages of state decision making.

So far there is no strong evidence on whether states are retarding new service expansion. According to a summary of a 1988 expert panel discussion: "It appeared from the discussion that while there was great concern about possible *future* impediments to the development of enhanced services that state regulation might erect, in practice the enhanced service providers (ESPs) are so new that there has not been time to see whether states actually will create significant obstructions" (Entman, 1988, p. 18). Although many new and potential information services are developing rapidly that could change this statement, it is largely still accurate today.

A third normative argument against state regulation is that there may be economies of scale in government's *production* of regulatory policy; 51 jurisdictions may waste substantial resources and one federal regulator might be cheaper and more effective than 51. The U.S. Department of Commerce's National Telecommunications and Information Administration estimated that direct telecommunications regulatory costs were about $1 billion in 1985, less than 1% of industry revenues, but that indirect costs of regulation were much higher. Quantitative evidence from the 50 states points to some economies of scale in developing analytically informed policy about rate structure (see Teske, 1990). Small state public utility commission staffs may not have the capacity to regulate thoroughly; some states have already recognized this by cooperating on a regional basis, to deal with regional holding company (RHC) level telecommunications issues.

Most of the authors of the chapters in this book focus on whether these theoretical concerns actually are relevant in American regulatory practice. Such an assessment inevitably involves judgments and trade-offs, on which experts will disagree. Does being closer to constituents really help improve policy, or does it prevent changes in rates that would enhance economic efficiency? Clearly Egan and Wenders disagree with Megdal on the answer to this question in their chapters.

Other relevant questions that are addressed within include: Are the states really innovating or just being dragged along by exogenous changes in technology and markets? Are state regulators inherently and inevitably preoccupied with local residential rates to the exclusion of all other important factors? Can states with small regulatory staffs really do an adequate job regulating?

POSITIVE EXPLANATION OF STATE REGULATION

The theoretical debates about the appropriate levels at which we *should* regulate may not be as important in actual regulatory politics as explanations focusing on why we *do* regulate at two levels. In reality, states do not only make regulatory policies because of arguments that they should, but because specific interests favor their policies and involvement. In analyzing positive explanations of state regulation, we consider which interests are favored by state regulators and how they might be harmed if the system were changed. Positive explanations require analysis of bureaucratic incentives and the incentives of interest groups. As with the normative evaluation, this positive evaluation is complicated.

Residential consumers have been protected by state regulation and seem to want it to remain in place. The FCC, particularly under the Reagan administration, was committed to raising their local rates (while cutting their toll rates) far more than are most state regulators. Consumers may feel that slower (than federal) state deregulation moderates change in an era marked by a transition away from the protection and stability of "Ma Bell."

Of course, residential consumers (who are also voters) were not directly active participants in many of the most important regulatory decisions in telecommunications. Historically, Gabel notes, they were not clamoring for state regulation in 1910, while Megdal notes that they did not favor AT&T being dismantled in 1982. No opinion poll that I am aware of has shown residential consumers favoring further deregulation of telecommunications services. A December 1989 *Washington Post* survey found that more consumers felt divestiture was a bad idea rather than a good idea, and that they remained confused about new services and aware of higher rates. This contrasts with an extremely favorable public opinion about telecommunications services prior to divestiture (Coll, 1986).

What other interest groups favor two-tiered regulation? Local exchange companies—the Baby Bells, GTE, and hundreds of other smaller firms—prefer state regulation when it maintains healthy profits, by protecting and nurturing them, and by keeping competitors out, or at least placing competition on a level playing field. The specific preference of a local company depends largely on their reading of the politicians and bureaucrats in the state, and their comparative advantage in different markets (see Teske, 1990). Some local companies have tried to maximize opportunities by bargaining across federal and state rulemakers, hoping to be allowed into prohibited markets in exchange for allowing competitors into their domains.

Congressmen seem to favor state regulation as it may take some political heat off them. Consequently, Congress has not moved to preempt state regulation. The 1980s were not the first time that Congress and state regulators have worked together; they blocked the FCC from reducing the subsidy from toll prices to local prices in 1970. According to Noll (1986): "Initially the FCC's criterion was to deregulate whenever competition could be relied upon to control AT&T's

market power, but by the mid-1980s the primary criterion has apparently become political feasibility: to deregulate as fast as possible without causing a pro-regulatory backlash in Congress" (p. 13).

The oversight relationship between Congress and the FCC, on the one hand, and state legislatures and state regulators, on the other, is interesting and opposite in some ways. In the 1980s, the FCC pushed deregulation and rate structure changes, whereas Congress resisted, sending signals without actually passing significant legislation (Ferejohn & Shipan, 1989). Although state regulators show great variety, as a whole they have deregulated far less than the FCC, and in several cases state legislatures have passed laws to push regulators to increase the pace of deregulation. Why did some state legislatures pass procompetitive deregulatory legislation when Congress did not pass any significant telecommunications legislation, and leaned in the opposite direction?

I find three reasons for this phenomenon. First, less than half of the state legislatures passed these deregulatory bills. Second, a majority of the bills were passed in states served by U S West, whose economic development clout and aggressive political pressure was a strong factor that state legislators were more likely to be concerned about than regulators (Teske, 1990). Third, in a few states, laws were required simply to update public utility enabling legislation that was 75 years old and that did not anticipate deregulation. Thus, on closer inspection, and exempting the U S West states, state legislatures have not been strongly out of step with state regulators and Congress. I examine this issue in more detail in chapter 4.

In considering the incentives and preferences of critical actors, we can not ignore the regulators themselves. Regulators and their staffs may want to keep their jobs and bias their decisions in favor of continued state regulation. Niskanen (1971) and many others chronicled the ways in which bureaucrats fight to maintain and expand their budgets. Indeed, in contrast to deregulation at the federal level in airlines (the CAB is out of business), trucking, and railroads (the ICC staff has declined by more than 50%), state telecommunications regulatory bureaucracies generally have *grown* since the AT&T divestiture, with the increased need for analysis and monitoring. In these other industries, federal bureaucrats helped to start deregulating themselves out of jobs (Derthick & Quirk, 1985). This has not happened often in state telecommunications regulation.

This explanation does not focus on a large number of bureaucrats; the typical state telecommunications regulatory staff is less than 20 people. However, these bureaucrats hold power more from control over information delivered to decision makers than from their numbers and a bias in favor of continued regulation could hardly be considered surprising.

With so many powerful advocates—Congress, many LECS, residential consumers and state regulators—we should not be surprised that state regulation remains in place. There are some opponents, however. As noted earlier, some LECs that are not regulated favorably by their states are opponents. Many interexchange carriers (IXCs) and ESPs dislike the possibilities of regulatory balkanization and increased costs. Some large business users of telecommunications also prefer a single set of federal regulations and favor the pricing policies of the FCC. Lobbying in 50 states and dealing with 50 different sets of rules is expensive and cumbersome for all of these groups, especially if they feel that state regulators and legislators are essentially captured by the LECs on issues to which they are opposed.

Powerful interests take their stands on both sides of this debate, and state regulation is also aided by its historical entrenchment. Regulation is often harder to remove than to initiate. States regulated intrastate railroads, trucking, and airlines for several decades after federal regulators were centrally involved. We are only one full decade away from the AT&T divestiture. State regulation also remains important in telecommunications because more than half of all revenue is generated at the intrastate level.

Because important interests want it to remain, there are few signs that state regulators will soon close up shop. The tradition, range, and power of state regulation, as seen in other industries, suggests that it will not soon wither away in telecommunications even if we argue that it should. As Noll (quoted in Entman, 1989) noted: "I believe that state regulation probably will have a role, at least for the next decade or two, but not because I believe that a major state role is either inevitable or desirable" (p. 17). And, as Doug Jones, director of the National Regulatory Research Institute, notes herein: "The decline and fall of state regulation has been prematurely implied or predicted a number of times by practitioners as well as academics." Even as we consider and recommend changes in the federal–state relationship, we should keep this point in mind.

THIS BOOK'S APPROACH

The book is organized to focus mainly on the political economy of state regulation in the 1990s. We also include serious looks backward and forward to gain temporal perspective. Although much of the analysis is positive, asking the question of what is happening in the states, we not only review and assess the current situation but we also consider normative concerns and make recommendations about how to improve telecommunications policy. Essentially we want to provide answers to the following questions: How are states regulating telecommunications in the brave new world of global markets, fiber optics, and digital technology? Do states vary significantly in their regulatory models? How are the politics of state and federal regulation different? Would a different federal–state relationship better serve national telecommunications goals in the future?

The chapter contributors have a wide variety of experiences and perspectives from which to tackle these questions. Although most have an academic background, two authors were themselves state regulators, three held significant federal policymaking positions in telecommunications, and several have provided practical consulting advice to business and government policymakers. We have also tried to present a mix of perspectives from the disciplines of economics, political science, communications research, and law.

In the following chapter, Gabel refutes the widely held supposition that states did little of importance in telecommunications before divestiture, except to decide on the rates of return on investment to which the firms were entitled (see also Cohen, 1992, for a complete history of state regulation). In fact, many important political-economic battles were fought at the state (and municipal) level, especially early in this century. Gabel discusses how states regulated in the early part of this century when there was competition, uncertainty, and the development of new paradigms of regulation, as is the case now. He clarifies some lessons that today's regulators should learn from this period.

Gabel notes that there was never a strong public demand for state regulation, but that the push came from business users of the growing telephone network and AT&T itself. Gabel suggests that whereas jurisdictional authority is not necessarily crucial, regulatory diversity is important to overcome information asymmetries that favor the regulated firms at the expense of the regulators. Gabel suggests that the burden of proof should not be on state regulation to prove its value but rather on federal regulation advocates to show how centralized national regulation could do a better job.

In chapter 3, Cole focuses on the issue of current state telecommunications regulation as a policy laboratory. With so much change, Cole notes that one of the most difficult tasks is simply to track what is happening in 50 states (plus the District of Columbia). Rather than trying to update an overly complicated 51 state by 99 policy matrix, as often seen in industry analyses, Cole addresses how the states have differed on a handful of the most significant issues, including competition or ease of entry, rate structures relative to costs, the pricing freedom given incumbent firms, residential consumer protection, service monitoring, and information infrastructure.

In chapter 4, Teske and Bhattacharya consider the policy role of governors, state legislatures, and economic development agencies, in theory and in actual practice. This chapter addresses the fact that, parallel to the development of so many new industry competitors and firms in telecommunications, state government actors beyond the traditional PUC regulators have started to play significant roles in telecommunications policy, especially in the area of infrastructure and economic development. They consider whether or not such activity is appropriate according to the original justifications for establishing expert PUCs. The chapter explores how legislative choices differ from regulatory decisions, and considers the role that economic development and infrastructure play in state regulation. Overall, they argue that the benefits of having additional government actors involved outweigh the costs and that economic development competition is spurring positive regulatory innovation by the states.

Chapters 5 and 6 address the costs and benefits of ongoing state regulation, respectively. The chapter on costs, by Egan and Wenders, argues that state regulation introduces significant efficiency costs that could be greatly reduced by a deregulation-minded FCC with preemption authority. They attempt to put values on the welfare losses associated with inefficient regulatory practices. They also recognize that it is not feasible to remove state regulation in the near term but that it should be a target down the road.

The benefits portion of this debate, presented in chapter 6 by Megdal, argues that state regulation has provided a laboratory and a compelling case against state regulation cannot yet be made. She notes that responses to diversity in tastes, competition between jurisdictions, and policy innovation result from the laboratory of state regulation. These elements can help achieve her four major regulatory goals of: (a) consumer protection from monopoly, (b) efficient regulation, (c) encouragement of competition, and (d) reasonable administrative burdens. She considers how a single federal regulatory apparatus might work and argues that it would not clearly be an improvement.

Three contributors comment on this debate. Jones offers an institutional defense of state regulation. Jones supports Megdal's balanced approach over the Egan and Wenders argument that all state regulation is obstructionist. Jones focuses more on institutional issues and political ideas than on a more narrow economic evaluation. He argues that the Reagan and Bush administrations, facing an ideological conflict, chose to focus on advancing deregulation with preemption rather than leaving decision-making power with the states.

Haring narrows down the economic justifications for preemption but leaves flexibility for future policymakers. He argues that shared powers may be appropriate as both federal and state approaches have costs and benefits. He proposes an "extrajurisdictional test" based on FCC and court review for when the federal government should have the right to preempt the states. He also argues that capture by the regulated firm may be more likely at the state level.

Gormley provides alternative perspectives on this debate that favor the advantages of state regulation. Gormley suggests that the economist's metric of efficiency should not be the only criterion by which to evaluate regulation at the state versus the federal level. He also suggests that the politics and bureaucratic decision-making processes are likely to differ at the two levels, with federal regulation influenced more by Congress and the states subject to closer consumer scrutiny. Gormley suggests that because our society has not yet reached consensus on both issues of fact and preferences in telecommunications, there is no pressing reason to jump to a single national policy. Gormley cites examples of the federal government learning from state regulation such as the California telecommunications lifeline program.

Finally, in chapter 7, Noam looks to the likely role of state regulation in the future. Noam presents a legal and political analysis of the history of federal–state interaction in telecommunications regulation and shows critical changes in recent years. He assesses the extent to which the important telecommunications policy issues in the next decade are appropriate for only federal action or policy shared with the states. Noam considers technological and political trends that may reduce the role of state regulation. Specifically, Noam addresses how the relationship with federal regulators should be restructured through legislation. He considers changes in the 1934 Communications Act that would be needed to improve the operation of federalism in telecommunications regulation, although he recognizes that such a comprehensive approach to this conflictual problem is unlikely to come out of Congress.

Two contributors comment on Noam's ideas. Geller agrees with Noam's analysis and argues that a new era of cooperation is the best that can be hoped for in the likely absence of Congressional action to improve the federal–state regulatory process. He expresses concern about some states' desire to regulate new information services and services like Open Network Architecture (ONA).

Tobias analyzes the actions of Congress in 1993 in preempting state regulation of cellular and wireless service rates. He considers whether or not this indicates a new willingness by Congress to tackle telecommunications federalism issues. He also analyzes the Clinton administration's new NII initiative and its implications for state regulation in the short and long run.

I synthesize and present concluding comments in chapter 8 and argue that as global telecommunications issues become more critical in telecommunications, battles between state and federal regulators may obscure more pressing competitiveness issues. As Noll (quoted in Entman, 1988) noted:

> The existing jurisdictional separation arises from the grossly outdated Communications Act of 1934, written when almost all of the revenues in the telecommunications industry came from local telephone service, and before such innovations as television, satellites, microwave, and computers. The notion that there is a meaningful technical and economic distinction between federal and state services was always a fiction, but it has become increasingly so. This fiction is likely to become ever more costly as the federal government withdraws from regulation, thereby increasing the capability of the states to impose mutually inconsistent requirements on what amounts to a national network providing national information services using equipment manufactured for a national market. (p. 17)

If anything, Noll's statement would be made more powerful by looking at many of these issues in the context of international telecommunications markets. In examining the following chapters, I urge the reader to keep in mind the larger picture of rapidly changing technology and markets, in contrast with a fairly stable regulatory structure that served well in the past, but may need revisions in the future.

2

FEDERALISM: AN HISTORICAL PERSPECTIVE

David Gabel
Queens College
City University of New York

The purpose of this chapter is to explore the historical development of federalism in the telecommunications industry. There have been two significant periods during which legislative bodies passed laws that preempted the regulatory authority of lower districts. First, beginning about 1907, states took over some of the regulatory responsibilities that had previously been held by the municipalities. The next major change occurred in 1934, when the U.S. Congress established the FCC.[1]

The first period is of special interest because it coincided with three developments that have modern parallels: the emergence of competition between telecommunication suppliers, the growth of nonexchange services made possible by new technologies, and the call by the Bell Operating Companies (BOCs) for a consolidation of regulatory authority.

The establishment of the FCC did not coincide with any major structural changes in the industry. The formation of the Commission is largely attributable to macroeconomic developments. Although there was clear concern about the earnings of the Long Lines division of AT&T,[2] the more important factors were the market failures that had led to the Great Depression and a political mood that questioned the propriety of allowing one firm to hold so much power.[3] The

---

[1]The Mann–Elkins Act of 1910 gave the ICC the authority to regulate interstate telecommunications. State authority was not preempted by the passage of the act.

[2]During the period 1913 to 1935, the average annual rate of return on net-book investment was 10.9% for Long Lines and 6.72% for the BOCs. Federal Communications Accounting Department, "Long Lines Department: Financial and Operating Summary," April 15, 1936, p. 15.

[3]Federal Communications Commission Telephone Rate and Research Department, "Final Report of the Telephone Rate and Research Department," p. 3, June 15, 1938.

Depression spurred government officials to increase oversight of this and other industries.

Because the first era of increased control by the central government offers more parallels and insight into today's policy issues, I focus on that time period.

In many ways preemption appeared to be a natural evolution. New production processes, such as the load coil and metallic loops, extended the distance and nature of communication links, and with these developments, different regulatory needs emerged. The states' assumption of what had previously been considered local government duties seemed to be driven by technology and corporate organization. As the use of long-distance lines increased during the first decade of the 20th century, city governments were the only existing regulatory bodies. However, they were not well positioned to monitor the growth of this new service, because the city councils' authority did not extend beyond the city boundaries. Consequently it seemed natural to assign telecommunications regulation to a body that had more extensive legislative power (Erickson, 1915).

State preemption of municipal regulation also seemed to be a sensible adjustment to the changing corporate structure of the nation's leading supplier—the American Telephone and Telegraph Company. At the turn of the century, for many reasons, including the need to improve the coordination of long-distance service, AT&T merged together some of its operating companies (Garnet, 1985). For example, prior to 1909, seven BOCs served New York state. Separate corporate entities were serving the state's larger cities—such as New York and Buffalo. But during that year, operations were consolidated into one firm, New York Telephone. At that point, it seemed to make little sense to leave the regulation of Bell to local authorities.

The centralization of control was also part of a larger political movement. The proponents of state regulation argued that preemption was a progressive reform because it removed power from corrupt, local political machines.[4] In addition, some leaders of the progressive movement, especially those associated with the National Civic Federation, believed that a cooperative working relationship should be established between the government and big business (Weinstein, 1968). The state governments, more than the municipalities, had the resources to carry out this cooperative effort. A well-trained, aggressive staff was needed to monitor the utilities.[5] Whereas large cities might have been able to assume

---

[4]McDonald, p. 118–120; *Milwaukee Journal*, October 17, 1906; *Milwaukee Daily News*, October 2, 1906; and "Arguments by J. A. Aylward," 1907, Wisconsin Legislative Reference Bureau.

[5]The Wisconsin Railroad Commission was the first state agency with the power to establish comprehensive regulation of the telephone industry. John R. Commons was one of the primary authors of the bill that established the Commission. Commons was concerned about the necessary conditions for effective regulation. He believed that although an active Commission required broad legislative authority, the record would be determined to a greater extent by the initiative and ability of the agency's personnel. Commons/La Follette, undated report located in the January 1905 paper of Robert La Follette, Wisconsin State Historical Library, pp. 1–6. In this letter, Commons was discussing regulation of the railroad companies.

this cost, smaller ones did not have the resources or general need to hire a permanent, professional staff.

RIVALRY AT THE TURN OF THE CENTURY

In my discussion of the reasons for the establishment of state regulation, one factor was noticeably absent—public demand for state intervention. During the period 1894–1913, AT&T faced stiff competition in many sections of the country. AT&T's rivals, known as the Independents, began by establishing telephone service in small towns and cities, as well as in the rural United States. AT&T had not developed these markets.

An important sector of the Independent's customers desired toll connections with the nation's larger cities. Businesses wanted to be able to contact their wholesalers, and some farmers wanted to have improved access to city markets.

At first, AT&T refused to interconnect with the Independents. In order to satisfy their customers' desire for toll connections, the entrants petitioned city governments for the right to construct exchanges in AT&T's primary markets— the nation's larger cities. These requests received considerable local support in markets in which Bell's customers were dissatisfied with the quality of telephone service, rates, or the lack of connections to surrounding communities.[6]

Most cities granted petitions for competitive exchanges. Head-to-head competition in such places as Cleveland, Los Angeles, Seattle, Philadelphia, Indianapolis, Buffalo, and St. Louis spurred market and technological developments. This competition compelled firms to actively seek new customers, and to offer high-quality service.

By 1901, in order to sustain its control over the large-city markets, AT&T concluded that it had to change its policy toward the less densely populated areas.[7] During the monopoly era (1876–1893), the market outside large and medium-sized cities had not been developed because AT&T believed that the marginal efficiency of capital was higher in large cities.[8] Competition taught AT&T an important lesson—due to the externalities of telephone service, some markets should be operated at an apparent loss. Although the direct revenues from serving some rural customers was less than the direct cost, it still made sense to develop these areas. Otherwise, if the regions outside the large cities

---

[6]See, for example, L. N. Whitney, "Report on Conditions in Indiana," August 1907, box 11, Museum of Independent Telephony.

[7]For example, see Fish/Pettingill, April 21, 1902, Presidential Letter Books (hereafter PLB), v.23-1; Fish/Davis, September 25, 1901, PLB v.16; and remarks of Charles Cutler, President of New York Telephone, at "Conference Held at Boston, January 23, and 24, 1900: Telephone Service and Charges" (hereafter "Telephone Service Conference"), pp. 226, 249, box 185-02-03, American Telephone and Telegraph Company Corporate Archive (hereafter ATTCA).

[8]*Wisconsin Telephone News*, 1 (December 1906), p. 1.

were left to the Independents, this would increase the demand for a second supplier in its profitable, large-city markets.[9] In order to protect its more desirable service territories, AT&T decided that it must compete with the Independents in rural areas surrounding the nation's cities.

Where direct competition existed, as well as in areas in which there were no major barriers to entry, there was essentially no public demand for state rate regulation. As the theory of contestable markets suggests, it does not take a large number of firms to obtain the positive attributes associated with competitive industry structure (Baumol, Panzar, and Willig, 1982). Duopoly rivalry, or the mere threat of rivalry, drove down prices, compelled the incumbent to improve its service, and led to a rapid growth in the expansion of telephone service to more households.

AT&T monopoly prices were also restrained by demand-side factors. Most important, outside large cities, the price and income elasticity of demand were high. Telephone service had not yet become "essential," and therefore there was little opportunity for a monopolist to earn supranormal profits. Because of this demand-side constraint, there was little need for state regulation in the vast majority of the nation's cities and towns.[10]

A strong sentiment for public control existed in the nation's largest cities, particularly in New York and Chicago. Competition was impeded in these metropolises due to municipal regulations intended to avoid further congestion. The rules required the placement of utility lines underground.[11] Because of the disruption that the construction of a second exchange would cause to the city's streets and commercial interests, municipal officials were reluctant to issue a franchise to the Independents.

This disruption could have been avoided by requiring the potential entrant to rent conduit and pole space from the incumbent, but the Independents were

---

[9]Remarks of President Cutler of New York Telephone, American Telephone and Telegraph Company, "Conference Held at Boston, January 23, and 24, 1900: Telephone Service and Charges," ATTCA, p. 226; Vail/Winsor, March 26, 1909, "Proposed Consolidation," ATTCA box 47.

[10]For example, when the State of New York held legislative hearings on the need to establish telephone rate-regulation, there were no public appearances at the Upstate public hearings. *Telephony, 18* (November 27, 1909), p. 561.

Competition was strong in Kansas, and the income levels constrained the prices that the telephone companies could charge. An opponent of regulating the telephone industry in Kansas remarked in 1909, that, "Not only is there no public demand, but there is no public *need* for this utilities' bill." "Argument of J. W. Gleed to the Kansas Judiciary Committees," p. 4.

In Wisconsin, although the public made frequent requests to their governors for protection from the Railroad monopolies, there was no call for state control of the telephone industry. See the collected papers of Governors Robert M. La Follette and James O. Davidson, Wisconsin State Historical Society.

[11]See, for example, New York Laws of 1884, ch. 534; and New York Laws of 1885, ch. 499.

concerned with the threat of monopolistic pricing by the owner of the structures, and the uncertainty of the availability of space to handle future growth.[12]

Typically, entry was opposed by the business and merchant's associations that represented the city's largest firms. Because the competing telephone systems did not interconnect, businesses often rented lines from both systems. The large businesses frequently opposed competition in the hope that they could avoid the expense of renting from the second supplier.

Residential customers were much less likely to rent telephone lines from both systems. Indeed such customers were likely to benefit from the significant price reductions that resulted from competition. Whichever supplier had the largest number of residential customers would also receive the greatest support from the business community. Consequently, the telephone companies promoted residential subscription in order to raise the value, and hence the price, to business customers.[13]

The large businesses were far from satisfied with the rates AT&T charged; monopoly pricing seemed to lead to supranormal earnings for the company in markets in which entry was not a constraint. Due to the absence of competition, the business trade associations asked the government to provide protection. In their appeals to legislators, these associations were particularly bothered by AT&T's profits and its rate structure.[14] They wanted to see rates lowered in the large cities and expressed no support for AT&T's rate philosophy that called for higher rates in urban centers in order to sponsor development in rural communities. Consequently in their requests for municipal or state control, these businesses were interested in seeing the establishment of a procedure that would limit AT&T's earnings in their cities to a "fair" rate of return.[15]

Municipal regulation posed a problem for AT&T. Chicago and New York City accounted for approximately 35% of the total earnings of the BOCs at that time. The rate of return in these cities was on the order of 13% to 16%, well

---

[12]*New York Tribune*, March 15, 1905; and Federal Communications Commission Accounting Department, "Report on American Telephone and Telegraph Company Corporate and Financial History," Special Investigation Docket No. 1, volume III, appendix 14.

[13]"Telephone Service Conference," footnote 9, pp. 168–169 (Sabin, President of Pacific Telephone), and 196 (U. N. Bethell, General Manager, New York Telephone).

[14]Federal Communications Commission Accounting Department, "American Telephone and Telegraph Company Corporate and Financial History," vol. 2, schedule 44 sheet 1 and schedule 46 sheet 4, January 16, 1937; Hearing Before Assembly Committee on General Laws on Assemblyman Krulewitch's Bill Entitled 'An Act to Regulate the Toll Charges for Local Telephone Communication,' " Albany, April 4, 1906, pp. 36–42; "Speech of Simon Sterne, Esq., Before the Assembly Committee on General Laws, January 30, 1889, in Favor of Bill Limiting Telephone Charges" (New York: George F. Nesbitt & Co., 1889); and *Telephony*, 17 (January 2, 1909), p. 22.

[15]Although AT&T challenged the claimed level of earnings, they did offer rate adjustments that limited earnings to a "fair" level. *New York Herald*, May 26, 1906; and U. N. Bethell, "Argument Before the Finance Committee of the Senate of the State of New York," March 10, 1898, Senate Bill No. 360: Telephone Rates (n.d., n.p.).

above the 5% cost of capital.[16] If municipal regulation prevented AT&T from earning supranormal profits in these markets, it would retard their effort to develop rural exchanges. It would also limit the available funds for their aggressive response to competition in other cities.

AT&T agreed with the business community that regulation was preferable to competition. In order to thwart the demand for competitive exchanges in the nation's largest cities, and also to address the grievances of its business customers, AT&T began to advocate the establishment of state regulatory commissions in 1906. The firm was the leading proponent of state regulation in Wisconsin—the first state to establish comprehensive regulation of the industry.[17] AT&T sought regulation as a means of preventing entry into its most profitable market, Milwaukee. State regulation was more desirable than municipal control, because in this way the large earnings in Milwaukee could be hidden by the losses of competitive exchanges.[18] Because the statewide earnings of Wisconsin Telephone were normal, AT&T expected that state regulation would have little effect on its pricing. In exchange for opening its books to state auditors, AT&T expected the legislature to prohibit the Independents from establishing a competitive exchange in Milwaukee.[19]

In 1906 AT&T also advocated regulation as an alternative to competition in Chicago and New York. But because the New York and Chicago Bell Companies had essentially been limited to exchange service in these two cities, the initial rate regulation involved price control negotiations between the BOCs and these cities' officials. The need to obfuscate the New York City and Chicago earnings encouraged AT&T to consolidate the operations of these profitable BOCs with less profitable AT&T-owned subsidiaries.[20]

Although municipal regulation did pose a threat to the firm's pricing and competitive strategy, it also offered some immediate relief to the most pressing problem being faced by AT&T—the threat that the Independents would establish competitive exchanges in these profitable markets. With entry into the heart of AT&T's profit centers—Chicago and New York—blocked, it was difficult for the Independents to sustain their competitive effort (Gabel, 1994). Hence, by 1907, many Independents had concluded that competitive exchanges could not be

---

[16]Federal Communications Commission Accounting Department, "American Telephone and Telegraph Company Corporate and Financial History," vol. 2, schedule 44 sheet 1 and schedule 46 sheet 4, January 16, 1937; *Wall Street Journal*, April 2, 1906; and Harry P. Nichols, "Report of the Bureau of Franchises Upon the Application of the Atlantic Telephone Company," October 12, 1905, p. 22.

[17]*Wisconsin Joint Committee on Transportation: Hearings on Public Utilities Bill, Number 933, A*, May 21, 1907, Wisconsin State Historical Society.

[18]Fish/Burt, August 19, 1905, Presidential Letter Books, Vol. 40, ATTCA; Wisconsin Telephone Company, "Telephone Talks," no. 4, 8, and 13, 1906, Wisconsin State Historial Society.

[19]*Wisconsin Joint Committee on Transportation: Hearings on Public Utilities Bill, Number 933, A*, May 21, 1907, Wisconsin State Historical Society, p. 89.

[20]Vail/Winsor, March 26, 1909, "Proposed Consolidation," ATTCA, box 47.

established in AT&T's profitable monopoly markets. Therefore, they either sold their properties to the incumbent or joined its network through a sublicensee contract.[21]

Competition had forced a radical decline in the price of telephone service. Rivalry, along with secondary factors, had caused the revenue per station to fall from a level of $90 in 1894 to $43 in 1907 (FCC, 1939). As competition diminished around the nation, customers were concerned that these price reductions would be reversed. No longer could users count on competition, or the threat of it, to ensure that quality service would be available at reasonable rates. The diminution of rivalry led to an increase in the demand for regulation. Although the opportunity for the earning of supranormal profits existed essentially only in the large cities, state level regulation nevertheless became the accepted solution.

Neither the hearings nor the legislation that authorized the establishment of state public utility commissions (PUCs) provides much insight into the policies that the legislature wanted the PUCs to pursue. Although they were clearly concerned that rates be "fair," there was little guidance as to how these agencies should balance fairness with policies that promoted the state's infrastructure. Instead, as is often the case with the U.S. legislative process, broad authority was granted to the delegated agency. It is mainly in the actual practices and policies of the PUCs, rather than the enabling legislation, that the objectives and means of preemption became apparent.

REGULATORY PRACTICES

Bernstein (1955) argued that regulatory commissions go through four phases: gestation, youth, maturity, and old age. During its early years, a regulatory commission may regulate an industry aggressively. Although its *raison d'etre* is to protect the "public interest," over time the agency loses sight of this objective. Due to limitations imposed by state and federal courts, most state PUCs authorized to oversee the telephone industry did not go through this initial aggressive stage. Rate cases were conducted in 1910 in much the same way in which they are handled today. In the first phase of the hearing, the revenue requirement of the firm was determined. In the second stage, the rate structures and levels that would generate the previously established revenues were set.

The first step was cumbersome and costly because of the legal requirement to determine the "fair value" of the utilities' property. The courts said that the determined rates must compensate investors for providing funds to the utility.[22] This return on investment was intended to reflect the opportunity cost of investing

---

[21]Historians typically attribute the turn of events in 1907 to the ascendancy of Theodore Vail to the presidency of AT&T. The policy of entering into sublicense agreements with Independents had been well established by this time. It was the supply of Independents willing to become sublicensees that changed dramatically.

[22]See, for example, Michigan Railroad Commission, Re: Michigan State Telephone Company, PUR 1918C, p. 84.

funds in a utility. Prior to the Great Depression, it was the general legal standard that the return be calculated as a percentage of the fair value of the property. Fair value was defined in terms of the economic value, or replacement cost, of the property.[23]

The regulated utilities, as well as the state PUCs, found it difficult and time consuming to estimate the replacement cost of the property.[24] Its value was typically determined by taking a physical inventory of the firm's facilities. The quantities of property were multiplied by current costs and then summed up in order to determine the fair, or market value, of the property.

These appraisals were conducted in excruciating detail. In 1914, accountants working on an appraisal of New York Telephone's investments in New York City, foresaw the need to determine the market value of almost 25 million units.[25]

Due to the burden of establishing the fair value of the facilities, few resources were left to conduct studies of individual rate items.[26] Cost studies were rarely undertaken, but when they were, the degree of analysis was often equal to or superior to what is done today at the federal or state level. For example, shortly after 1910, the City Counsel of Chicago and the Wisconsin Railroad Commission authorized detailed studies of the cost of providing different services in Chicago and Milwaukee, respectively.[27]

In both jurisdictions, the analysts identified the fixed and variable costs of production. Fixed costs were allocated between classes of service based on the number of lines in service. Variable costs were assigned to the customer classes based on peak and 24-hour usage.

---

[23]*Smyth v. Ames*, 169 U.S. 466, 546-7. See, also, *Democratic Central Committee of the District of Columbus v. Washington Metropolitan Area Transit Commission*, 485 F.2d at 800-01.

[24]C. A. Wright and D. B. Judd, "Standardization of Telephone Rates" (Columbus: Ohio State University Engineering Experiment Station, 1923), p. 1; and *Final Report of the Joint Committee of the Senate and Assembly on Telephone and Telegraph Companies, State of New York* (Albany: J. B. Lyon Company, 1915), pp. 446–447.

[25]*Telephony*, Vol. 67, no. 4, 7/25/14, p. 30. In 1914, the state authorized $100,000 for the appraisal. It was expected that it would take 2 years to complete. Because of the cost and time requirements, most areas in the State received only cursory attention. *Final Report of the Joint Committee of the Senate and Assembly on Telephone and Telegraph Companies*, pp. 25–28. The $100,000 expenditure is approximately equal to $1.2 million in 1988 dollars.

The Minnesota Commission undertook a statewide valuation of Northwestern Bell Telephone's properties at a cost of $1,000,000. The study took 3 years to complete, and in the end, the Commission found that the fair value of the plant was approximately 4% in excess of the book cost of the properties. Nebraska State Railroad Commission, Re: Northwestern Bell Telephone Company, PUR 1923B, pp. 117–120.

[26]Thompson and Smith (1941, p. 208); and Federal Communications Commission, "Final Report of the Telephone Rate and Research Department," pp. 20–21, 43.

[27]William J. Hagenah, *Report on the Investigation of the Chicago Telephone Company: Submitted to the Committee on Gas, Oil and Electric Light* (Chicago: Henry O. Shepard Co., 1911); and Wisconsin Public Service Commission, "Valuation and Bases of Allocations and Apportioning Property Groups of the Various Classes of Service," 1915, box 10, series 41/4/8, Wisconsin State Historical Society.

There are three striking aspects of these cost studies. First, the quality work done at the state and city level was clearly superior to the analytical work that was being done by the federal regulatory agency—the ICC.[28] Second, in the markets in which there was the greatest need for regulation, such as Chicago, city officials were able to marshall the expertise to do the same high-quality work that was being done by the nation's preeminent regulatory agency—the Wisconsin Railroad Commission (Stehman, 1925). Third, the work done by these two regulatory bodies was also superior to the work effort made at most commissions.

Most PUCs' boards were overwhelmed by the task of determining the revenue requirement. Pricing decisions had to be based on information other than cost studies. The size of the agency budgets left few remaining resources for the undertaking of detailed cost-of-service studies.

Furthermore, there was little push on the part of the regulated firm for detailed cost analysis. AT&T's strategists recognized that depending on the nature of regulation, state oversight could be either harmful or helpful to its operations. Regulation was beneficial to AT&T where it blocked competitive entry. State oversight could hinder AT&T if it interfered with the firm's goals. AT&T wanted the PUCs' attention restricted mainly to determining the firm's revenue requirement.[29]

Although cost studies were an input to AT&T's internal rate setting process, the BOCs believed that cost studies should be kept out of the public record. Accountants and economists could estimate the cost of service, and this information would give the PUCs some independent indication of how rates should be designed. AT&T wanted the PUCs to base prices on the value of service.[30] This required little or no cost data, and gave the regulated firm the greatest pricing freedom. It was much easier for a PUC staff or its consultants to use engineering and accounting data to determine the cost of service, than to appraise

---

[28]The ICC was granted regulatory authority in 1910, but did little more than adopt AT&T's accounting practices as the standard for the industry, processed annual reports, and approved mergers of competing firms as being in the public interest (Brock, 1981, pp. 158–159; and Joint Application of Rock County and Wisconsin Telephone for Certificate that Acquisition Will Be in the Public Interest, 70 ICC 636).

The U.S. Congress and the staff of the FCC attributed the lack of activity at the ICC to inadequate funding (Federal Communications Commission, "Final Report of the Telephone Rate and Research Department," pp. 3–4). The ICC's large staff concentrated on railroad issues. The failure to establish a communications division within the Commission has been attributed to the lack of public interest in interstate telephone regulation (Wheat, 1938, pp. 846–847; Fainsod & Gordon, 1941, pp. 366–367).

[29]Stigler (1971) argued that a firm will favor regulation if the benefits derived (protection) exceed the burdens assumed (constraint on prices and earnings). AT&T apparently believed that if commissions did not undertake cost studies, price constraints would be, essentially, nonbinding, and this would increase the net benefits of regulation.

[30]See, for example, W. S. Ford, "Memorandum: Concerning Certain Peculiar Features of Telephone Exchange Service," September 10, 1901, "Telephone Rates—Basis—1880–1908," box 12, ATTCA; "Telephone Rates: A Few Suggestions as to How They Should Be Defended" (n.a.), "Rates—Basis for Determination—1906," box 48, ATTCA; and Lee (1913, p. 79).

the relative demand for different services. Therefore, due to the combination of lack of resources and staff initiative, and AT&T's rate philosophy, most of the early state regulatory decisions reflected the value-of-service pricing concept.

With few exceptions, the states decided not to consider the cost of providing service in each market; instead they established rate classifications that were based on the number of telephones in service. The number of exchange access lines was considered to be the best gauge of the value of service. Based on this criteria, customers in large cities were charged the highest prices. The earnings in these exchanges were then used to cover the apparent loss in localities in which the traffic could not bear compensatory rates (Baird, 1934).

CONCLUSION

Three primary conclusions can be reached from this history: jurisdictional authority is not necessarily crucial, policymakers should be mindful of the need for diverse approaches, and regulatory policy can be used to promote techno-logical change. State-level regulation was favored by AT&T in 1907, in part to obfuscate the price–cost relationship of various services. Firms can hide data on earning levels regardless of the regulatory jurisdiction. Because today's telephone companies provide multiple services in different states, and also offer products through unregulated operations, any agency will have a difficult time determining which costs are associated with which services.

A commission is effective to the extent to which it can establish service standards and quantify the cost of providing the multiple products offered by the exchange companies. In addition, regulators must develop a perception of the market. Knowledge of demand-side factors will help a commission set prices that meet criteria such as sustainability, optimality, and promotion of growth (e.g., developing new network services).

The pricing of new services is an area in which a new regulatory agenda must be considered. The pricing history of the telephone industry shows that new services typically have been offered at a loss. AT&T was willing to serve markets and sell products at a loss because it believed that once a critical mass was achieved, or certain externalities came to fruition, the service would eventually become profitable.[31] In today's contentious atmosphere, and especially because of the legitimate antitrust concerns of several interested parties, it would be risky

---

[31]In 1926 the President of AT&T, H. B. Thayer, remarked that "telephone service was not created to fill a demand . . . the service creates the demand. That is the business of our system, to try to discover and determine what it is that will be helpful to the people of the United States in the way of service and then to provide it. The demand follows the creation of the service instead of being impelled by it." *Proceedings of the Bell System Educational Conference for Faculty Representatives of Colleges of Liberal Arts and Collegiate Schools of Business* (New York: AT&T, 1926, p. 11). See, also, Hall/Vail, May 12, 1885, box 1011, ATTCA.

for a local exchange company to admit that a new product was being offered below cost initially.

Because new services often require a redesign of the existing network, it is necessary to establish a mechanism that will account for the full cost of introducing a new product or service. This means recognizing that it is quality, and not just quantity, that drives new capital expenditures.[32] Furthermore, a procedure must be developed in which the basic-usage telephone subscribers are compensated for the money that they have previously provided for the establishment of these new services (Gabel, 1990).

The measurement of the cost of service is an art that the FCC and the state PUCs have largely failed to master. The task is becoming more challenging as new services are increasingly based on software enhancements. Because telephone companies have the best sense of their markets, even where cost standards have been established, they essentially control the rate-setting process. This is part of the larger regulatory problem, as summarized by Platt: "The private sector has better access and use of legal, economic and technical information about essential public services" (Platt, 1989; p. 44).

This informational asymmetry exists at both the state and federal levels. The agencies' problems are compounded by the general uncertainty regarding the nature of the future demand for services.

The former Bell System had a wonderful way of dealing with uncertainty— new products and managerial innovations would be tested in selected cities, and subsequently analyzed by the parent company and the heads of the different operating companies. It was through this inductive learning process that the firm was able to develop its long-term strategy. Today, state PUCs share information and learn from one another, just as they did at the start of the decade. One clear loss resulting from federal preemption would be the latitude and initiative to test different policies. The FCC would have a difficult time authorizing and supervising regulations that varied across regions of the country.

History also suggests that before federal oversight be increased, we need to determine the reasons for inadequate state supervision. The lackluster record of the first state commissions was due mainly to the court-imposed requirement that the fair value of the property be the basis for rate setting. Replacing state with federal regulation would not have eliminated this obstruction during the pre-World War II era. The proponents of federal preemption must identify the limitations of state regulation, and show how the FCC will be able to overcome these barriers.

Recent history suggests that the FCC does not have the know-how or the initiative to handle effectively the difficult standards, pricing, and cost issues associated with the development of intelligent, broadband, and open networks.

---

[32]For example, data services need a "cleaner" line than the type of connection needed for voice services.

During the 1974 antitrust case, the Department of Justice argued that the FCC did not have the expertise to regulate the operations of AT&T (Noll & Owen, 1989). In the postdivestiture era, this essential point has not changed. Following the decision to apply price caps to local operating companies, the Commission's Chairman, Alfred C. Sikes, remarked to the *New York Times*: "I don't believe that career Government people, or for that matter career non-Government people, can find out what the true cost of a service should be."[33]

AT&T's original regulatory strategists would be quite happy to read Sikes' comment. They realized that as long as the regulatory commissions did not have data about costs that could be used to gauge the reasonableness of a rate, regulators would be seriously hampered in their efforts to challenge the company's rate proposals.[34] As noted earlier, AT&T did conduct internal cost studies; they just did not want the regulatory commissions to use cost data as an input to policy decisions.

An intelligent decision-making process, by either private or public policymakers, involves collecting and analyzing information. For internal purposes, the telephone companies have identified, and will continue to identify, the cost of providing different services. Instead of trying to develop the needed cost data for policy decisions, the FCC has selected quick, but inefficient cost and rate making solutions.

The record of the states is only somewhat better. But their response to this lack of cost data has been more constructive. Instead of claiming that they can not determine the cost of service, the state commissions have used their resources to develop some first-rate cost models.[35]

Finally, some argued that governmental controls of economic activity, such as regulatory commissions and antitrust laws, serve to offset free market decisions that might otherwise result from technological innovation. Where vocal sectors of the body politic begin to lose their advantage in the wake of technological

---

[33]September 20, 1990, p. D2.

[34]Sikes' comment is consistent with the Commissions' decision that it did not have the expertise to determine the cost of different services, and therefore would use relative-use as an allocator between regulated and nonregulated operations. "Separation of Costs of Regulated Telephone Service From Costs of Nonregulated Activities," 2 FCC RCD 1298 (1987), modified on reconsideration, 2 FCC Rcd 6283 (1987), modified on further reconsideration, 3 FCC Rcd 6701 (1988), petition for review pending Southwestern Bell Corp. v. FCC, D.C. Circuit No. 87-1764 (filed December 14, 1987).

In its modifications to the Separations' procedures, the Commission has reduced or eliminated its recognition of the cost difference between exchange and nonexchange services. The modifications are inconsistent with the internal cost studies done by the industry. For example, although every long-run incremental cost study recognizes that the switching costs of an interoffice call is higher than the cost of an intraoffice call, the Commission favors using relative minutes of use to allocate switching costs.

Because of the Commission's expressed inability to identify the costs of existing services, it will be in a poor position to judge the reasonableness of ONA and CEI tariffs.

[35]See, for example, Pollard (1990) and Mount-Campbell and Choueiki (1987).

change, they may lobby for nonmarket controls that mitigate or eliminate dislocations from technological change (Hughes, 1977; Owen & Braeutigam, 1980). The early history of the telephone industry shows how regulation may be used to aid the growth of the network. State regulation reduced the regulatory authority of municipal governments. This change in authority established the conditions for the flow of funds from the cities to cover the costs of extending the network into less populated areas. Federal regulation may serve a similar purpose—earnings from high-traffic volume, low-cost states may be used to cover the cost of network services in areas where the population density and traffic volumes are relatively low.

STATE POLICIES AND ACTORS

3

STATE POLICY LABORATORIES

Barry Cole
Columbia University
Michigan State University

As in earlier periods of regulatory turmoil, since the AT&T breakup state tele-communications regulators have promoted different policies and are thus acting as laboratories, at least in some respects. Based on their statutes and policy pronouncements, nearly all states seem to be aiming for similar goals, although some of these goals are vague and some are even internally contradictory. States have pursued varied approaches to achieving these goals.

The seven main goals of most states in telecommunications regulation today appear to be: (a) protecting consumers from monopoly abuse and from undesirable cross-subsidization of unregulated competitive activities; (b) promoting equity, by fostering universal service and by ensuring rural populations receive access to similar services and prices as their urban counterparts; (c) promoting compe-tition which will, in turn, lead to greater productivity, expanded service, lower prices, and more options; (d) enhancing economic efficiency by moving prices toward costs; (e) promoting innovation and efficient, technologically advanced telecommunications and information services; (f) maintaining high service quality standards; and, in recent years, (g) creating a telecommunications infrastructure that will aid in the economic development and competitiveness of the state.

As state policymakers try to achieve these goals, they find that many significant factors are beyond their control. Among the most important of these factors are technological change, changes in the general economy (e.g., inflation), and the development of competitive markets. But state policymakers do exercise signifi-cant control through their regulation of cost allocation, pricing, and depreciation practices for intrastate public network investment, which collectively represents about 80% of U.S. public telephone network investment.

REASONS FOR DIFFERENT STATE POLICIES

There are several reasons for the different state policies. The first is that state legislators have varied in the extent of legal authority they have delegated, explicitly or implicitly, to PUC regulators. State telecommunications laws vary in how deregulation is defined, whether it is mandated or not, what services are specified, what process requirements are required, what equity issues are included, and what oversight and sunset provisions are provided. The following chapter by Teske and Bhattacharya addresses in some detail the reasons for these legislative differences.

Second, states vary in the size and nature of existing and potential service markets and in their market environment. For example, Iowa has over 140 small telephone companies serving rural areas, whereas 8 states have 10 or fewer small telephone companies. Eleven states are single local access and transport areas (LATA) states,[1] and these have generally resisted facilities-based intraLATA competition to a greater degree than other states. Moreover, the amount of actual and potential competition within the states varies considerably.

Third, political pressure for change varies by state. Williams and Barnaby (1992) argued that 38% of state regulatory reform plans were initiated by the PUCs, 30% by state legislatures, and 24% by the telephone companies. Some legislators, such as Assemblywoman Gwenn Moore in California, have been very active; Moore introduced and helped enact more telecommunications bills than any other state legislator. In terms of telephone companies, U S West has been the most aggressive in trying to achieve deregulation statutes in all 14 states in which they operate. Pressure can also come from large business users of telecommunications services or from new competitors. Bypass of public network facilities has occurred most intensively in a small number of states.

Fourth, the regulators and regulatory environment varies in each state. Some differences relate to the preferences and priorities of individual commissioners and their key staff members. Teske (1990) showed that these views, as represented by the "regulatory climate" measures used by Wall Street analysts, do influence policy choices, and Cohen (1992) provided similar evidence of the impact of

---

[1]Local access and transport areas, or LATAs, were created and defined in the Modified Final Judgment for the divestiture of AT&T in 1982. They were assumed to generally fit local service regions. States were given the authority to decide whether or not to allow competition across LATAs within their states, as most states did fairly soon after divestiture, as well as whether to allow competition within LATAs, which was a more complicated choice that evolved more slowly in most states. See Teske (1990) for a political economy explanation of different state intraLATA competition choices.

regulators. In effect, we have 51 mini-FCCs, each staffed by different human beings who view similar problems differently. Thus, for example, when all states were faced with the problem of how to deal with an FCC order mandating open entry for customer-owned coin-operated telephones (COCOTs) within their state, state regulators adopted varying policies, including where the COCOTs could operate, how much they could charge customers, and what services they must provide (e.g., 911, directory and operator assistance, and coin return).

Regulatory differences are also affected by staff resources and expertise. In 1994, California and New York had over 50 professional staff members working exclusively on telecommunications.[2] On the other extreme, 5 states had none, and 17 others had 5 or fewer staff members. One Rocky Mountain state has traditionally resisted assigning even one person exclusively to telecommunications, on the grounds that if that person left the staff, they would have no expertise remaining.[3]

State policy is also sometimes influenced by the actions of other states. In a survey of all state PUCs by Hudson (1990), 78% thought decisions by other commissions in their regions were "somewhat important" to them, and 17% thought they were "very important." Only regulators in Alabama, Florida, Indiana, Nevada, and New Mexico answered that those decisions were "not important." In some regions, particularly in New England and the 14-state Mountain region served by U S West, the PUCs meet and share information on a regular basis.

EXAMPLES OF STATE POLICY CHOICES

Since the AT&T divestiture, states have faced many important policy issues, but have often responded in different ways, based on many of the factors cited earlier. The most contentious issue the states faced immediately after divestiture was probably rate design generally, and local residential rates in particular (see Teske, 1990, and Cohen, 1992, for how these were handled). Rate increase requests were at record levels in the mid-1980s. However, state regulators were fortunate that a variety of factors reduced rate increase requests in later years. Low inflation, low interest rates, productivity improvements in telecommunications technology, and changes in the 1986 Tax Reform Act became favorable to local rates. Some states, such as California, have consistently fought to keep local rates as low as possible, whereas others have recognized trade-offs and allowed rate increases in order to achieve other important goals.

Another particularly crucial and vexing policy issue for state regulators has been determining "effective competition." States have struggled to obtain relevant

---

[2]State PUCs also usually regulate energy and gas issues, and sometimes water, sewer, transportation, and insurance industries.

[3]Note that the tremendous variation in staff is not only true in telecommunications regulation. The California PUC has over 500 professional staff, whereas 10 states have 20 or less.

information to make more informed decisions about both local rates and effective competition. As Gabel noted in chapter 2, such information ideally would include: (a) detailed market analysis to identify what market segments are competitive, where LEC pricing flexibility is appropriate, and where entry may occur and undermine existing rate structures; (b) detailed knowledge of LEC economic costs to prevent predatory pricing, more carefully target cross-subsidies, and set cost-based rates; (c) information relevant to insuring LEC cost efficiency; (d) greater knowledge of market structure effects of technological change; and (e) sophisticated market monitoring to facilitate targeted responses by the state regulators. No state has been able to gather all of this information, but states with more PUC staff resources have been able to do more than others.

Cost and effective competition issues are particularly difficult in local markets. In major cities, competitor firms like Teleport and Metropolitan Fiber Systems (MFS) have tried to compete with the LECs, going after their biggest customers (see Teske & Gebosky, 1991, for details). To respond to that competition, the New York State PSC was at the forefront of state efforts to develop interconnection policies that would allow such competition to be fair and not preempted by the LECs. With similar entry pressure in Chicago, the Illinois PUC developed the idea of "Telecommunications Free Trade Zones," to open up local markets to competition, but also giving the LEC pricing flexibility to compete fairly for that traffic. As of July 1994, only New York, Illinois, Maryland, and Washington have approved full local competition, while Connecticut, Pennsylvania, Michigan, and Wisconsin have stated that they intend to authorize it.

States also face ongoing problems related to service to rural areas. Of course, some states have larger rural populations than others. Rural policy problems include possible increases in rural long distance rates if urban and rural costs are deaveraged, failure to benefit from competition because it develops very slowly, and less potential to expand and reduce costs of rural service through technical improvements (see Parker, Hudson, Dillman, & Roscoe, 1989).

A less central example, but one that is still salient to consumers, was provided by telemarketing policy in the early 1990s. Telemarketers had discovered that automated dialing devices were more efficient for them than human operators. With the expansion in usage, many consumers began to feel inundated with such calls and complained in large numbers to PUCs and to legislators. Because of consumer complaints, telemarketing became the leading telecommunications concern of state legislatures in 1991, accounting for 25 of the 123 telecommunications bills passed that year.

States handled the issue in different ways. For example, New Mexico required that messages only be given with the recipient's consent and that no telemarketing calls were allowed before 9:00 a.m. Indiana and Washington state allowed calls after 8:00 a.m. Florida required telemarketers to be licensed and to file scripts and sales literature with the state Agriculture and Consumer Services Department.

This led to complications for the telemarketing firms operating in a national market environment. Their pressure and consumer complaints pushed Congress to act on this issue. The House and the Senate passed bills requiring the FCC to develop rules limiting telemarketing and taking consumer preferences into account on this issue.

ALTERNATIVES TO RATE-OF-RETURN REGULATION

As discussed elsewhere in this book, problems have arisen with the traditional rate-of-return approach used for most regulated utilities, and states have, to varying degrees, experimented with alternatives. Proponents of changes suggest that over 30 states adopted new forms of regulation through 1993. Traditionalists suggest that in most cases these reforms are simply modifications to rate-of-return review. How radical have state innovations really been?

A decade after the 1984 implementation of divestiture, some states had done relatively little to effectuate regulatory change. The states often placed in this group include Indiana, Wyoming, Pennsylvania, North and South Carolina, and Washington, DC. Others have instituted more change, but significantly less than the LEC, and in some cases, even less than the state PUC desired. For example, in Illinois, the PUC issued a 1989 order to put a profit-sharing incentive regulation scheme into action. But in November 1991, an appellate court overturned the order and forced the PUC to reinstate traditional rate-of-return regulation on Illinois Bell. The Court ruled that the existing state law allowed no other option. That law was to sunset at the end of 1991, but attempts to enact a new law, one that would have authorized the implementation of alternative methods of regulation, failed and the sunset was extended until 1993.

In Michigan, the telecommunications statute was scheduled to sunset in 1992. Michigan Bell, the Telephone Association of Michigan, and the Communications Workers of America (CWA) pushed a major deregulation bill introduced in the state Senate, which would have created a new telecommunications regulatory body with jurisdiction over any form of two-way telecommunications. Under the bill, local service and access were to become nonprofit monopolies of the telephone companies. All other services were to be wide open to competitive entry. However, a coalition of small businesses, non-Bell competitive telecommunications providers (including the Michigan Cable TV Association and the Michigan Telemessaging Association), and various consumer groups opposed the bill, fearing that cross-subsidization and stringent certification standards for competitors in regulated services might deter small companies from entering regulated markets. The result was a law far less extreme than the original bill. It was subjected to 100 Michigan House amendments during its initial pass at the bill, 47 of which were adopted in whole and 15 in part by the committee of House and Senate conferees. The Michigan PUC was retained in the new law, with broad discretion to regulate services and authorize competition in regulated services. The PUC also retained authority to protect consumers and competitors from marketplace abuses, regulate accounting practices and quality of service, and investigate and resolve consumer complaints as well as conflicts between providers.

In those states in which some form of incentive regulatory plan was introduced, the plan was often offered to the telephone company on an optional basis. Usually the company responds positively. But sometimes the company finds the innovation unpalatable and chooses to remain under traditional rate of return regulation, as in Utah. U S West told the Utah PUC that the sharing formula in the plan (for earnings above a 12.2% rate of return) did not provide sufficient incentive, particularly with earnings capped at 17%.

In Colorado, U S West even wanted to reverse some deregulation that had already occured. The firm proposed reregulation of special access, Centrex, feature options, and any service introduced after 1987, such as voice mail and other enhanced services. U S West cited the cumbersome cost allocation procedures required by the law each time it wanted to introduce a new service or substantially modify a deregulated one, and noted that deregulated services accounted for only about 5% of gross revenues. U S West also claimed that bringing such deregulated services back under regulation would put the substantial time and resources devoted to cost allocation calculations to better use.

As these examples illustrate, it is quite difficult to indicate precisely the degree of innovation that states have provided, because characterizing and labeling what these states have done with respect to alternative forms of regulation is often arbitrary. Even the experts disagree, or at least use different terminology. This is illustrated by comparing summary reports prepared by Bell Atlantic, Bell South, and Southwestern Bell staff responsible for tracking policymaking at the state level. For example, 1990 reports individually prepared by all three firms provided almost identical detailed descriptions of "alternative" regulatory actions in Florida, Kentucky, Illinois, Washington, Michigan, Mississippi, and Minnesota. But after describing what was done, the Bell Atlantic summary labeled the alternative plan adopted in each state as "Rate Stability/Incentive Regulation," whereas Bell South called it "Earnings Sharing," and Southwestern Bell classified it as "Revenue Sharing." *State Telephone Regulatory Reports* called these state actions "Rate of Return Incentive" plans. One should therefore be wary of statements that *x* number of states have taken action *y*, as the categories may or may not accurately describe the innovation.

For example, among the terms used to describe various reforms, innovations, and modifications that have taken place at the state level are: *banded rate of return, banded pricing, pricing flexibility, detariffing, service specific detariffing, stepped regulation, rate of return incentive, price deregulation, complete deregulation, price caps, price regulation, social contract, negotiated social contract, earnings sharing, revenue sharing, profit sharing, incentive sharing, incentive regulation, rate stability, rate moratorium, rate stayout, rate case moratorium, rate of return elimination, rate equalization, rationalized regulation, flexible regulation*, and *rate innovations*, which may be *flexible, open, revenue neutral*, or *tiered*. Some of these terms are taken to mean the same thing. Most of the terms describe alternatives that are more flexible than rate-of-return regulation.

In formulating and reviewing alternative regulatory plans, state PUCs often consider the following factors: (a) the length of the plan, with most ranging from 2 to 5 years; (b) treatment of basic local service rates; (c) adjusting prices toward costs; (d) how to handle depreciation reserve deficiencies; (e) allowing pricing flexibility to meet competition; (f) incentives for innovation and cost efficiency; (g) treatment of exogenous factors such as taxes and FCC separations changes; and (h) financial and service quality reporting. Of course, these considerations often overlap. For example, after 1992 an increasing number of alternative plans were emphasizing service quality as an incentive for innovation.

In some states, the criteria to be used by the state PUCs to determine if an alternative plan should be adopted are spelled out in the relevant state law in some detail. In others they are left almost entirely to the PUC. For example, in Washington state, the statute requires the PUC to make written findings of fact as to each of seven policy goals (public interest, fair rates, service quality, etc.) in ruling on any proposed alternative regulatory plan. Before any plan may be adopted or modified, the PUC must make a positive finding on each of these goals.

Forty-one of the 50 states surveyed by Hudson (1990) noted that the impact of telecommunications on the states' social or economic development was a consideration in its decisions on regulatory alternatives. Those states that cited specific criteria for accessing these impacts referred to promoting efficient use of facilities, enhancement of state networks, infrastructure support for the state economy, impact on attracting and retaining industry, and rural access to the same services available to urban customers. In the next chapter Teske and Bhattacharya consider the impact of bringing these infrastructure criteria into the regulatory decision-making process.

MEASURING SUCCESS OR FAILURE OF STATE POLICY EXPERIMENTS

Measuring and evaluating the *potential* success of proposed new regulatory approaches is extremely problematic. But measuring and evaluating the *actual* success of an alternative plan, once it is in place, is also not easy, as illustrated by an independent auditor evaluating Alabama's 1986 Rate Stabilization and Equalization (RSE) Plan, a new approach to regulating South Central Bell. The auditor, Theodore Barry & Associates (1991) noted that even though the plan was merely a modification of an RSE Plan previously adopted to regulate both the Alabama power and gas industries, "The RSE process is a relatively unique concept in the area of regulatory oversight. As such, very few models exist by which any commission can make an assessment of the program with a high level of confidence about its conclusions. The normal standards of comparative cost/benefit used in most such evaluations have not yet been formulated and will not be for some time" (p. 5). The firm was thus forced to establish a set of parameters within which a systematic analysis could be conducted. In explaining the criteria it used in evaluating the success of RSE, the auditors noted: "If a decade of experience in such matters were available to each party, or it a dozen other commissions had comparable programs, it could be concluded that sufficient knowledge exists to warrant stronger definition" (p. 11).

Mueller's (1993) evalution of the impacts of Nebraska's deregulation bill is the most careful comparative analysis of state policy experimentation. Despite rate deregulation, he found local rate increases within reason, but intraLATA toll rates that are higher in Nebraska, in part because of lack of competition allowed in the bill. With U S West hoping to achieve regulatory gains in other states, the firm may have kept rate increases in Nebraska lower to generate goodwill; thus the experiment and its evaluation may not be properly controlled.

Some opponents of state alternative forms of regulation argue that the firms' financial results (profits) become the performance index that measures its behavior. Thus, they argue, the reward and performance index are essentially the same things—profits—and the incentive becomes circular.

The Aspen Institute Telecommunications Regulation conference (summarized by Entman, 1988) attempted to develop a consensus of state goals and strategies, as well a comprehensive set of measures to gauge success. Those measures include: penetration rates, basic service rates, customer complaints, cross-subsidization, quality of service, productivity increases, investment planning, costs of regulation itself, deployment of new services, community perceptions of the industry, economic growth, complaints from competitor firms, extent of competition, price changes, diffusion rate of new technologies, rural service, and industry rates of return. Unfortunately, any such long list includes many measures that may involve trade-offs, depending on the particular regulatory policy pursued.

Better evaluation of the state policy experiments is a clear priority. Such evaluation should be performed by government and academic researchers, rather than only the interested party firms, to reduce actual bias and the perception of bias.

VARIED STATE APPROACHES TO THE INFORMATION SUPERHIGHWAY

Unless the Clinton administration's NII initiative, backed by Vice President Gore, takes preemption further than most expect, the 50 states will be critical players in the development and evolution of an "information superhighway" or intelligent telecommunications network. As states retain policy responsibility for cost allocation, pricing, and depreciation practices for 80% of the national *public* network investment, their decisions about the price of access, rural issues, interconnection rules, privacy concerns, and cross-subsidies will be important. Ultimately, the most important technological aspects of these advanced networks concern the integration of multiple services over the customer access line into the home or business, which is primarily under the jurisdiction of the states, a jurisdiction that would be the most difficult for the federal government to preempt.

State opposition to preemption on this issue was illustrated by their reactions in 1991 to the National Broadband Development Act, which aimed to mandate nationwide (to the home) deployment of broadband fiber optic networks by 2015 and require the states to develop a plan for FCC approval. Through NARUC, the states declared their opposition to preemption of intrastate network investment choices. That bill did not pass Congress, but state PUCs did get the opportunity to register with Congress their opposition to extreme preemption under the justification of a national information superhighway.

Even as the direction of development of an information superhighway becomes clearer, there are still many uncertain elements. Although policy experimentation by innovative states is valuable, most state policymakers recognize that it is equally unwise to recklessly lead the charge for the information superhighway, and risk falling off the cliff, as it would be to go to the other extreme of sticking its head in the sand and refusing to accept future trends that will affect their state.

States have employed at least five general approaches to modernizing networks to build the information superhighway. These five are not necessarily mutually exclusive; some states have combined elements.

The first is the legislative approach. State law can be a vehicle for promoting infrastructure modernization and fostering an evolution to an intelligent network. Recent laws in New Jersey and Illinois are illustrative. The introduction of the New Jersey bill followed a consultant's report. It includes a goal to wire the state with fiber optics to develop a state-of-the-art telecommunications infrastructure for economic development. Rate stability is ensured by the plan over much of the investment, particularly for the most basic local service options. Similarly, Illinois' law is based on two principles: protection of captive consumers and flexible regulation for competitive services that will encourage infrastructure investment.

Second is the task force approach. The Governor's Telecommunications Task Force in Michigan released a report in 1990 called "Connections: A Strategy for Michigan's Future Through Telecommunications." The Task Force was co-chaired by then Governor Blanchard's wife Janet. Assisted by 220 representatives of business, government, and education, the task force made 53 recommendations, including the establishment of a Cabinet Council on Information Technology that would develop a state-of-the-art network for state government and incentives for private sector involvement in a modern network. Governor Blanchard was defeated in the 1990 election and many of these recommendations are not being implemented by his successor. However, the Michigan experience showed that the task force approach is viable for other states.

Pilot networks are a third approach. In 1991, the Missouri PUC issued a report called "Network Modernization and Incentive Regulation" that argued that the "most effective way to inform the public of the need to modernize the network is by example" (p. 3). It called for a task force to establish a pilot educational and health-care broadband network. An increasing number of other states are considering or actually creating pilot networks to help stir interest in telecommunications modernization and to better understand the realities of operating and programming broadband systems. It is important to develop some consensus about the specific goals of the project (e.g., cost efficiency or new services demonstration) and the means to evalute whether or not these goals are actually met.

A telecommunications master plan respresents a fourth approach. With a significant push from then-Senator Al Gore, the Tennessee PUC in 1990 adopted a detailed $400 million master plan for accelerated telecommunications technology deployment throughout the state over the next decade. After committing to the plan, the local exchange telephone companies are able to operate under an alternative regulatory framework, which includes extended earnings review and some pricing flexibility. The extent to which new technology and new services, such as ISDN, will be available in Tennessee is significantly accelerated with this master plan. The plan has been controversial and has not been fully implemented as originally designed. Nevertheless, it provides an example of a detailed modernization plan based on an exhaustive analysis of the current telecommunications infrastructure that other states can imitate.

The fifth approach is creating a comprehensive public database to be used in planning and policymaking. New York state began this process in 1991. The database is expanded and updated continuously to include relevant information related to user needs by various segments (residential, business, mobile, etc.), modernization alternatives and plans, evaluation tests along the way, and secondary benefits to the state.

One recent approach that combines elements of the task force, pilot network, master plan, and database approaches is the New York State Telecommunications Exchange. Through an ongoing process, the Exchange hopes to integrate regulatory and economic development initiatives and use state government as a positive force to develop the telecommunications industry in socially desirable directions, still relying on market forces for most of the investment and service provision.

The shape and scope of the information superhighway or intelligent network within a given state will ultimately be greatly affected by the presence or absence of public policy. In beginning to navigate new, uncharted waters, some states are adopting some or all five of the approaches just outlined, to meet their own needs, visions, institutions, traditions, and political realities.

CONCLUSIONS

States have acted as policy laboratories in the decade since the breakup of AT&T. Public utility commissioners have tried a number of approaches to telecommunications regulation, as influenced by their own analysis and political input from governors, legislators, telephone firms, new competitors, large business users, and consumers.

Sometimes analysts overstate the actual variance in state policies, as many different terms are used to describe similar or related policies. But there are real differences between state approaches, and some have already acted as models not only for other states, but for the federal government. There are also great differences in states' capacities to handle these issues.

In the important issues related to the information superhighway that lie ahead, states will play a critical role, unless, of course, much of that role is preempted by the federal government. Even in the event of preemption that would go far beyond today's two-tiered system, states are likely to be left with a some policymaking role.

As states consider that role, more actors, beyond the PUCs, are laying a claim to telecommunications turf and expertise. The next chapter analyzes the implications of additional state actors entering the arena of telecommunications policy.

4

STATE GOVERNMENT ACTORS
BEYOND THE REGULATORS

Paul Teske
Mallika Bhattacharya
State University of New York
at Stony Brook

Contributors to this book are addressing the question of whether there are already too many actors on two different levels of American government—national and state—who collectively are fragmenting telecommunications regulation. It is important, therefore, to assess the role of actors beyond the public utility commission regulators at the state level. We believe that the active involvement in telecommunications policy of other state political actors—governors, legislators, task forces, and economic development agencies—stimulates greater policy experimentation than the PUCs alone would implement.

In many states, the question of whether legislators, governors, or economic development agencies *should* be involved has become moot; they are already playing an important role in telecommunications. The critical questions are: What have they done? What should they be doing in the future?

Politicians have become attracted to telecommunications, as it promises productivity, high-technology economic development, and a comparative advantage in the information economy. Many of their business constituents recognize the increasing importance of the telecommunications infrastructure.[1]

---

[1]It is true that many industry stakeholders try to use these other actors and the economic development potential of telecommunications to gain more favorable regulatory policies than they might get from the PUC. Although this second set of "rent-seeking" activities can distort policy, attempting to link self-interest to the public good is a common practice in our American political system with multiple forums for policy, checks and balances, and political pluralism.

WHY STATES HAVE PUCs

Before examining other actors, it is important to remember why and how PUCs
were established. They are a compromise between two ideas, expert policymaking
and political accountability. Society needs regulators who have expertise and the
political insulation to make difficult decisions, if necessary. However, society
does not want to have completely insulated regulators who are not accountable
to larger political influences.

PUC regulators do have the *expertise* and the resources that other actors,
especially legislators and governors, do not have. The politicians delegate power
to the PUC bureaucracy, but provide a general framework in law about how to
progress and what decision-making procedures to utilize. PUC bureaucrats must
base their decisions on facts and law, or face being overturned in court. Also,
legislators can pass new laws if they do not like the direction in which bureaucrats
are interpreting laws, and, at least in theory, they can reward and punish bureaucratic behavior with future budgets (although some PUC budgets come out of
utility receipts rather than from legislative authorizations). Similarly, governors
appoint the PUC regulators in most states, which presumably gives governors
some influence or at least knowledge of what to expect from regulators.

Consequently, according to this model, about three fourths of our state regulatory commissioners are appointed by governors and approved by state senates,
and the remaining one fourth of states elect their commissioners directly, for a
more direct form of political accountability. We might expect this difference in
direct versus indirect political accountability to affect policy choices, but a wide
range of research is largely inconclusive on this issue.[2]

Recent scholarship points out that legislators do not always delegate for reasons of expertise, but often for *political-electoral reasons*. Legislators (and governors) sometimes prefer *not* to make difficult political choices that can harm
influential interest groups and constituents, so they pass the "hot potato" issue
on to bureaucrats. These politicians are assuming that voters will be less likely
to blame them for harmful choices made by bureaucrats, particularly if these
choices are made incrementally over time and if they involve a high degree of
technical detail and complexity (as with much of telecommunications regulation;
see Gormley, 1986). Certainly this has been true for Congress and telecommunications policy over the past several decades, as they have not been able to pass
significant legislation with so many powerful interests lobbying them in different
directions.

Delegating to bureaucrats is not an *unconstrained* strategy. At least sophisticated interest groups and some voters will realize that legislators passed the hot
potato and may expect their elected officials to solve the problems that are created

---

[2]See Teske (1990) and especially Cohen (1992) for a contingent theory and more detailed study
of elections and state telecommunications regulators.

by subsequent bureaucratic choices. Still, most PUC regulatory choices in tele-communications are not overturned by the politicians, and state regulators seem to make most of the important decisions.

WHY ARE OTHER ACTORS INVOLVED?

Since state telecommunications re-emerged as an important policy area after the agreement to breakup AT&T in 1982, many state legislatures and governors have approved laws to restructure regulation and to allow deregulation. In some cases, their action was necessary, as public utility enabling laws in these states were half a century old and did not anticipate the need for flexibility or deregulation. Some state politicians have taken this opportunity to pass laws that go beyond simply giving regulators authority to deregulate, but instead prescribe more de-tailed regulatory practices, as in Illinois and Florida, for example.

However, governors and legislators have not limited their role in telecommu-nications policy to regulation. Some have seized on telecommunications as *the* crucial infrastructure for state economic development in the information economy. As Cole noted in chapter 3, in several states economic development agencies or task forces also have become involved in telecommunications policy. As the U.S. Congressional Office of Technology Assessment (1990) noted: "Also steering the States in diverse directions is the fact that many State officials are now beginning to recognize the economic development potential of telecommunica-tions" (p. 364).

As the American economy has shifted to information-intensive services and as the international competitiveness of our firms has become an increasing con-cern, policymakers have looked to telecommunications as a vital infrastructure for economic growth. Although the common analogy of telecommunications networks as the highways of the 21st century is far too simplistic, it does have a compelling logic that has galvanized policymakers. Vice President Al Gore claims to have coined the term "information superhighway" and it has become part of common usage.

Evidence from numerous studies illustrates that telecommunications is a *nec-essary condition* for economic development, though probably not a sufficient one. As with any single factor in economic growth, it is difficult to isolate the precise impact of telecommunications. See, for example, Wilson and Teske (1990), New York City Partnership (1990), Parker et al. (1989), Schmandt, Wil-liams, and Wilson (1989), Coopers and Lybrand (1989), Smilor, Kozmetsky, and Gibson (1988), Arnheim (1988), Coalition of Northeastern Governors (1987), Moss (1986, 1987), Hanneman (1986), Blazar (1985), Saunders, Warford, and Wellenius (1983), and Hardy (1980).

Even if advanced telecommunications networks and services had no positive effect on economic growth or productivity, a position that even most skeptics would suggest is extreme, as long as policymakers think that voters believe in this linkage, states will pursue such policies. And, in the multistate competition for jobs, each state faces prisoner's dilemma incentives that can force them to compete even if they do not see a link, because competitor states are taking such actions.

Thus, a new issue has emerged at the state level. *Should* these other policy actors (governors, legislators, and economic development agencies) step aside and let the regulators use their own expertise to make telecommunications policy within the traditional public utility regulation framework or do they have a legitimate role to play, particularly in the modernization of infrastructure for economic development? To provide some perspective on this question, we assess what these other actors have done in actual practice.

WHAT HAS THEIR INVOLVEMENT MEANT?

This section assesses the role played by new laws, gubernatorial task forces, and development agencies in state telecommunications policy. How do policy choices by these actors differ from PUC choices and how do they influence the PUC regulators?

Legislatures

Between 1982 and 1992, 29 states passed important bills affecting state telecommunications regulation. Some state legislatures, including Washington, North Dakota, Minnesota, Utah, Mississippi, and Florida passed more than one major telecommunications law in this period. Thus, at least in 29 states, legislatures and governors have been directly involved in the telecommunications regulatory policy since divestiture.

As Table 4.1 illustrates, 21 state laws explicitly allow their respective utility commissions to deregulate or modify regulation on a service-by-service basis, often specifying three categories to be considered: competitive, partially competitive, and noncompetitive. Some of these laws are more explicit than others about which services fall in which category. Another eight state laws are enabling statutes that give authority to the PUCs to deregulate or to develop alternative rate making plans, but do not specify how or when to do so. Seven state laws mandate the study of alternatives to rate-of-return regulation. Another four state laws, led by Vermont in 1987, can be classified as oriented toward social contracts, where prices for basic services are held down and network modernization is ensured in return for more regulatory flexibility for the local telephone companies. Nebraska's famous 1986 law stands in a class by itself that could be

TABLE 4.1
State Deregulation Laws

Service by Service Statute	Modified Service by Service Statute	Enabling Statute	Social Contract Legislation	Study of Alternatives to Regulation
Iowa	Colorado	Connecticut	Vermont	Indiana
Arizona	Minnesota	Mississippi	Idaho	Florida
South Carolina	Missouri	Ohio	North Dakota	Utah
Indiana	South Dakota	Washington	Minnesota	Colorado
Mississippi	Nevada	North Carolina		Nevada
Montana		Florida		North Carolina
Georgia		Utah		Washington
Nevada		West Virginia		
Texas				
New Mexico				
Wisconsin				
North Dakota				
Oregon				
Utah				
Michigan				
Washington				
Illinois				

Source: Authors' analysis of state laws.

labeled radical deregulation by legislative fiat (see Mueller, 1993, for an analysis of Nebraska's law).

To try to understand these laws, and the motivations behind them, we need to examine past research. Some scholars believe that, because legislators face elections and most state regulators do not (because about three fourths are appointed), legislative decisions will be oriented more toward short-term political incentives than bureaucratic choices would be. We might also expect different political parties to develop different telecommunications policies, based on the interests of their core supporters. Although empirical evidence from other areas of state regulation is mixed, in telecommunications there is evidence that legislators do influence policy and that party control is important and in the expected direction. Teske (1990), analyzing policy choices made from 1984–1987, found that state legislatures controlled by Republicans were significantly more likely to favor higher local rates and lower toll rates, but less likely to favor intraLATA competition than legislatures controlled by Democrats. Similarly, Cohen (1992) found that, from 1977–1985, Democratic majorities in state legislatures led to larger cross-subsidies in local flat rates from business to residential consumers.

When we analyze state legislative actions, it is important to note that nearly half of the laws passed were in the 14 states served by the most aggressive Baby

Bell company, U S West. U S West had an explicit strategy to pass legislation and "bypass" the regulators (and for good political reasons, see Teske, 1990). Therefore, a large percentage of these bills resulted from strong pressure from a powerful interest group on relatively willing (and Republican) legislatures that are among the nation's *least* professional, by most measures of salary, staff, resources, and time in session.

That such strong political pressure can be effective is not surprising. Hudson's (1990) survey found that state legislative staff expertise in telecommunications was limited; 27 states had no professionals working on telecommunications, most had one or two, and the maximum number was 6. In part, this was because no state had a legislative committee devoted exclusively to telecommunications, but usually as a part of utility or commerce committee. Hudson also found that legislators rely heavily on telephone lobbyists and PUC staff for their telecommunications expertise.

To further understand legislative motivation and delegation of authority to PUCS, we performed an in-depth comparison of all 50 state laws. As illustrated in the Appendix, we measured legislative involvement in two ways—legislative delegation of authority to PUC bureaucrats and oversight provisions in the legislation. Some state legislatures wrote telecommunication laws that defined in very specific terms what the agency was supposed to do under specific conditions. Other state legislatures passed laws that defined in vague terms what the agency was supposed to do, leaving more freedom for PUC regulators to interpret and enforce the laws. Furthermore, legislators can delegate authority, but retain power by oversight provisions. Some legislatures have done this, whereas others have not included strict oversight of the PUC in their laws.

In Table 4.2 we present a cross-tabulation of the specificity of scope of telecommunication laws. Very often state legislatures write vague telecommunications laws that are open to different interpretations. In the category of the regulation laws, approximately one third of the states (18 out of 50) have laws that are not specific about the role of the PUC regulators. In the category of the deregulation laws also, again about one third of the states (10 out of 29) do not clearly define bureaucratic scope.

TABLE 4.2
Specificity of Scope of Regulation and Deregulation Laws

		Number of Regulation Laws	Number of Deregulation Laws
Laws Present	No Authority	0	0
	Very Specific	13	6
	Medium Specific	19	13
	Not Specific	18	10
	Total	50	29
Laws Absent		0	21
Total States		50	50

Looking at instruments of regulation, we find that 38 state legislatures have given many regulatory instruments to their regulators.[3] None of the state legislatures that have written deregulation laws have left their PUC with few regulatory instruments, and most instruments are retained to regulate local telecommunication services. Thus, there is much less variation in legislative control over the bureaucrats in this area.

The oversight requirements in the state telecommunication laws are skewed in the direction of being very stringent. About half of the laws provide for strict oversight requirements for their respective PUCs. Still, there is considerable variation in the oversight requirements across the various state legislatures because the remaining states are evenly split between medium and relaxed oversight requirements.

Now, we address the question of why legislatures have been involved in telecommunications in different ways. Political science theory suggests that the concepts of uncertainty and conflict are critical. As uncertainty increases, legislators have more difficulty in deciding the best policy for their state and for their own political futures. McCubbins (1985) argued that under increasing uncertainty, legislatures will try to be less involved, and delegate broader authority, so that if something goes wrong in the future, the bureaucrats can be blamed for it. They expect broader delegation of authority to be linked with more strict oversight provisions, to constrain bureaucrats.

Others, like Moe (1989), argued that with increasing uncertainty, the existing legislative majority will try to insulate the agency from future interference from the current political opposition (who might later be in power) by writing more specific laws with less delegation of authority and less oversight provisions.

We analyze the impact of both political and economic uncertainty on legislative activity.[4] By using a multiple regression methodology, we find that political uncertainty, as measured by turnover rates, had statistically significant negative effects on both the scope of the law (see Table 4.3) and oversight requirements (see Table 4.5), thus confirming Moe's expectations. Economic uncertainty also had statistically significant negative effects on both the scope of the law (Table 4.3) and oversight requirements (Table 4.5), also confirmimg Moe's expectation.[5]

Conflict is the second factor that political scientists expect to affect legislative involvement. As Congress has shown, if the interests of legislators and their constituents conflict greatly, passing a law will be difficult. McCubbins (1985)

---

[3]Here, it is difficult to split up the set of laws into regulation and deregulation categories simply because 29 states have both regulation and deregulation laws, and although the regulatory instruments usually come from the regulation laws, few of them are described in the deregulation laws.

[4]We measure political uncertainty by the turnover rates in the state legislatures and economic uncertainty by the average change in personal income over the 5 years before the passage of the law.

[5]The coefficient for economic uncertainty is positive due to the specific measure used, but the relationship is as predicted by Moe.

TABLE 4.3
Effect of Independent Variables on Specificity of Scope of Law

	Variable	Coefficient	Standard Error	t Statistic
Uncertainty	Income	3.4528e–03	2.01124e–03	1.71660**
	Turnover	–4.00619e–02	2.07828e–02	–1.92765**
Conflict	Conflict1	–0.56177	0.83370	–0.67383
	Party	–9.48720e–02	0.66669	–0.14230*
	USWest	–1.64909	0.79523	–2.07373***
	Loop	1.59617e–02	9.22300e–03	1.76974**

$R^2 = 0.393$
$N = 48$
*Significant at 90% level; **Significant at 95% level; ***Significant at 99% level.

argued that with conflict, legislators delegate more to bureaucrats, forcing them to make the hard choices. He also expected a tightening of the oversight requirements to ensure that the bureaucracy does not abuse this power. On the other hand, Moe (1989) expected that because policy involves compromises between the majority and the minority groups, the majority will have to yield to the opposition's demands for higher delegation and less strict oversight to some extent. As a result, McCubbins and Moe expect the similar delegation behavior but different oversight requirements under conflict.

We examine three different conflict variables. One captures the conflict between business telecommunications users and consumer groups in each state.[6] The second captures conflict between urban and rural consumers, who face dif-

---

[6]Business power in a given state is measured from the number of headquarters of Fortune Service 450 firms in that state. These Fortune Service 450 firms represent the industries with largest average toll usage. Firm headquarters are used because they are the places from where the largest amount of telecommunication traffic flows, and because they carry political clout in the state where their headquarters are located. If the number of headquarters of Fortune 450 firms in a particular state is above the average for all the states (in this case, it works out to be 8), then business power is considered to be high in that particular state. Conversely, if it falls below that average, then business power is considered low in that particular state.

According to Teske (1990), lobbyists representing the widely dispersed small consumers have become increasingly active since 1980. These consumer advocates lobby for groups like the elderly, poor households, and rural members. Gormley (1983) categorized the level of activity of such grassroots consumer groups in the late 1970s across the 50 states as either high or low. Consumers are also represented by advocates funded by state governments. Another measure for the consumer power in any given state can be the level of activity of these government-funded consumer advocates in that state. Gormley also categorized the government-funded consumer advocacy in each state as either high or low. We used Gormley's categorization as a measure of consumer power in any state. If both business and consumer groups are strong in a state, we assume that there is high conflict. Any other combination of business and consumer power is considered as low conflict.

TABLE 4.4
Effect of Independent Variables on Regulatory Instruments

	Variable	Coefficient	Standard Error	t Statistic
Uncertainty	Economy	1.33225e–03	1.13374e–03	1.17510
	Turnover	–1.05517e–02	1.17153e–02	–0.90068
Conflict	Conflict1	0.3181	0.4699	0.67697
	Party	–0.1301	0.3758	–0.34638
	USWest	0.4692	0.4482	1.04689
	Loop	8.40313e–03	5.08590e–03	1.65224*

$R^2 = 0.196$
$N = 48$
*Significant at 95% level.

ferent local telecommunications costs.[7] The third measure captures conflict related to U S West's aggressive behavior in its 14 states.[8]

Again, by using multiple regression methodology, we find that the presence of U S West has a significant effect on the specificity of scope of the law (Table

---

[7]Urban consumers have to subsidize the rural consumers so that the rates for the rural consumer do not become astronomical because of the relatively high access costs of reaching remote parts of the country. This conflict can be measured by the percentage of the total population of a state living in metropolitan areas. The hypothesis is that the lower the percentage, the greater is the possibility of conflict because the cost of subsidization will be higher and the relatively small urban population would not like to subsidize the vast rural population. On the other hand, the higher this percentage, the lower will be the possibility of conflict because it will be an accepted value in states with a large urban population that the urban majority will have to subsidize the rural minority. Another measure would be the access loop costs. Higher access loop costs indicate higher percentage of rural population and consequently, higher conflict. But these two are highly correlated, so using any one is sufficient.

[8]Sometimes a single organization or interest can be so overwhelmingly powerful in affecting policies in the states in which it is present, that it can be labeled as the dominant interest in those states. Its presence itself will determine the level of conflict in the state. One such powerful actor in the telecommunication arena is U S West. As Teske (1991a, 1991b) pointed out, it pursued a very aggressive political strategy. It went directly to the legislators and got passed the legislation of its own choice, thereby bypassing the regulators altogether. He also pointed out that the 14 U S West–controlled states were fundamentally different from the remaining states in many respects. It was serving states that are sparsely populated with few large cities. U S West states had few headquarters of large companies, which use telecommunications intensively and have political clout to influence government policies. On the other hand, the consumer advocacy in the regulatory proceedings was also consistently lower in the U S West states than the rest of the country. In other words, U S West was relatively free of pressure from the large business users, as well as consumer groups, and was the sole interest group dominating the political scene in these states. Therefore, we assume that the U S West states are low on the conflict scale, whereas the rest are considered high.

TABLE 4.5
Effect of Independent Variables on Oversight Requirements

	Variable	Coefficient	Standard Error	t Statistic
Uncertainty	Economy	4.84234e–03	2.34376e–03	2.06606**
	Turnover	–6.84186e–02	2.44298e–02	–2.80062**
Conflict	Conflict1	–2.49185	0.94358	–2.64084**
	Party	–0.45524	0.75060	–0.60650
	USWest	0.49505	0.94087	0.52616
	Loop	8.17465e–03	1.05361e–02	0.77587
Delegation	Scope	–0.32084	0.17579	–1.82525*

$R^2 = 0.292$
$N = 48$
*Significant at 95% level; **Significant at 99% level.

4.3).[9] This finding confirms anecdotal evidence that U S West went directly to state legislatures to get detailed laws that would minimize PUC regulators' discretion and power.

We also find that rural–urban conflict has a significant positive effect on both specificity of scope of law (Table 4.3) and regulatory instruments (Table 4.4) provided in the law, confirming McCubbins' and Moe's expectations. Business users and consumer conflict has a significant impact only the oversight requirements (Table 4.5), in the direction expected by Moe.

In addition to uncertainty and conflict, we also measured the impact of political party differences between the branches of state government. This brings the governor into the decision-making process, which we also explore more in the next section. We expect that if the same party controls both houses of the legislature and the governor's office, these politicians will trust each other more, and legislators will delegate more authority to bureaucrats, some of whom are appointed by the governor. We find that party differences reduce the specificity of the scope of the law (see Table 4.3).

Thus, the evidence is clear that legislators take political factors related to uncertainty and conflict into account when they delegate power to PUC regulators. They show concern for their re-election prospects and they show that legislative action is a compromise. The powers they do and do not give to regulators no doubt influence the policies that regulators develop, as Cole illustrated in chapter 3.

Governors and Economic Development Agencies or Task Forces

The previous analysis shows indirect evidence that the governor's political party influences telecommunications regulatory legislation. In addition to legislation,

---

[9]U S West did not, however, have a significant impact on the number of regulatory instruments or on oversight provisions.

several governors have played an active role in commissioning major reports, organizing economic development task forces, or establishing new departments within economic development agencies to deal with telecommunications issues outside of the public utility commissions.

State activity of this type is increasing rapidly. Williams and Barnaby (1992) surveyed state telecommunications economic development activities. They found that, as of mid-1992, 16 states (of 51) had developed infrastructure modernization plans, 14 had implemented infrastructure studies, 6 had held task forces,[10] and 9 had developed explicit quality standards. We analyze briefly these activities in a few major states.

As early as 1986 the New York State Department of Economic Development funded a major consulting study of telecommunications economic development, regulatory and tax issues, which was independent of, but included input from, PUC regulators.[11] In his 1992 State of the State message, Governor Mario Cuomo called for establishment of a "Telecommunications Exchange," which led to a 37-member task force. The Exchange's report called for regulatory changes, including moving toward an open "network of networks," a new universal service funding mechansim, a level regulatory playing field putting CATV under PUC regulation, and benchmarks for modern infrastructure. It also called for economic development applications, including the strategic use of government networks, the promotion of telecommunications services diffusion, and the establishment of a new Office of Telecommunications within the Department of Economic Development to manage and promote these efforts.

For many years, California was known for considerable consumer-oriented telecommunications legislation, particularly from Assemblywoman Moore in lifeline services (see Jacobson, 1989). Harris (1988a) wrote a telecommunications report for the California Economic Development Commission and Pacific Bell established its Intelligent Network Task Force in 1987 to bring social and economic issues related to the network to the forefront. Still, regulatory change in California lagged behind many other states. With the 1990s economic decline in California, telecommunications issues have taken on a new urgency. Governor Pete Wilson's 1993 State of the State address challenged the PUC to use telecommunications to improve the economy and a 1994 plan will aim toward more competition in the next few years to advance new services.

These large states have not been the only ones to link telecommunications and economic development. In addition to its radical deregulatory legislation, in Nebraska former Governor Bob Kerrey made the Center for Telecommunications and Information a division of the state Department of Economic Development

---

[10]The composition of these task forces varies by state, depending on who sets it up and for what purpose, but they often include members of the telephone industry, state government officials, consumer advocates, and representatives of the disabled, elderly, and minorities.

[11]The most important immediate motivation for the study was to deal with a telecommunications equipment/real property tax issue that had implications for New York businesses.

in 1986 and emphasized telecommunications-based development, a strategy that his successor, Governor Orr, also followed.

Maine Governor Brennan's Task Force in 1985, based in the state Planning Office, took aggressive steps to focus on economic development and the telecommunications infrastructure, as well as examining more traditional regulatory issues. The state has offered free access to its own rights-of-way as an incentive for carriers to extend fiber optic networks. State Planning Office Director Richard Silkman argued that "public utility commissions generally do not have the statutory authority to consider rural economic development when developing rate structures. That authority belongs with state legislatures" (quoted in Parker et al., 1989, p. 31).[12]

In Michigan, former Governor Blanchard assembled a task force, largely outside the PUC, to determine what could be done to upgrade the state's telecommunications infrastructure to stimulate economic development. Their report proposed linking the state together with a broadband network and leveraging state agencies and other private actors to get them involved in telecommunications. Blanchard's successor has been less active in telecommunications.

Although many of these tasks forces were established to address economic development issues that PUCs seemed to be ignoring, not all PUCs are uninterested in this issues. Members of PUCs in states with large cities, like New York and Illinois, and some regulators in very rural states, have been concerned and aggressive about economic development issues. Hudson's (1990) survey showed that *most* utility regulators see their role as telecommunications policymakers rather than simply as focusing on narrow regulatory issues, and this view of proactive regulatory approaches largely is confirmed by the telephone associations in each state. Still, as Hudson (1990) noted, "PUCs may be seriously understaffed, and may have little awareness of development issues" (p. 49).

HOW SHOULD STATE POLICYMAKING BE STRUCTURED?

Should states adopt the federal telecommunications policymaking model? That model includes the traditional three branches of government—the President, Congress, and Judge Greene—as well as two agencies, the FCC and the National Telecommunications and Information Administration (NTIA). Of course, the FCC is analogous to the state PUCs; it is *the* regulatory agency for interstate communications and has a parallel (though much broader, including broadcasting and

---

[12]And, "Silkman pointed out that public utility commissions could play a role in telecommunications and economic development," but "we have abrogated responsibility for making them focus on that." For example, when the Maine PUC had a $3 million windfall to dispense—more than the legislature ever spends for economic development—it chose merely to lower residential rates by 50 cents a month. And, "My only point . . . is that PUCs are making their decisions with no input from legislatures arguing for rural economic development" (Parker et al., 1989, p. 33).

media) role. FCC regulators, like most PUC members, are appointed by the executive and approved by legislators, and presumably are responsive to both (Ferejohn & Shipan, 1989, show their responsiveness to Congress).

The NTIA is part of the Department of Commerce. NTIA was established by President Carter's Executive Order 12046 in 1978 to provide for the coordination of the telecommunications activities of the Executive branch, and was previously called the Office of Telecommunications Policy, established within the President's office by Nixon in 1970 (Geller, 1989). NTIA advocates economic development for telecommunications and focuses on our international competitiveness in this area, by issuing industry analyses, defense concerns, and assisting with international trade negotiations.

Relations between the FCC and NTIA have not always been smooth. In its *Telecom 2000* (NTIA, 1988) report NTIA argued that FCC decisions should be overruled by the executive branch on issues of "overriding national security, foreign policy, international trade, or economic policy" (p. 20), a potentially broad set of areas. Similarly, the FCC and Congress have been concerned about an overly imperialistic role by the NTIA, although this has largely not materialized.

In comparing states to the federal model, of course, Congress is analogous to the state legislatures. The most obvious point about Congress and telecommunications in recent years is that it has not been active in passing laws. Congress, especially its subcommittee members, continually sends important signals to regulatory agencies, which have been heeded by the FCC (Ferejohn & Shipan, 1989). The telecommunications interest group environment (see Berry, 1989) is one of the most complex, which makes it extremely difficult for Congress to legislate without causing substantial harm to one or another powerful group.

The federal level actor that does *not* have a parallel at the state level is Judge Greene and the Department of Justice regulatory apparatus related to the divestiture, nor are we suggesting that their should there be such a role at the state level. Although activity by Congress, the FCC, and NTIA can make policy coordination difficult, it is the additional influence of Judge Greene that most critics find problematic in the current regulatory regime in Washington.

We believe that it is appropriate to have these other actors involved at the state level. The changes in telecommunications markets and policy after 1984 have been substantial, not incremental. PUCs are often not given a mandate to consider economic development and perhaps they should not be given it. Regulators set rates and monitor service quality, balancing the financial health of producers with the desire to keep consumer rates low. Such choices certainly *affect* economic development. They affect economic efficiency directly through rate structures, and they affect the future network through allowing or disallowing specific investments and through depreciation policy, even if the concept of economic and infrastructure development is not an explicit factor in many of these decisions.

If the main political incentive PUCs have is to maintain low local rates to the detriment of economic efficiency (Noll, 1986), a matter of some controversy, but something we have some evidence for (Teske, 1990), then economic development pressure from other actors may stimulate innovative regulatory policy. Interstate competition for mobile firms can help stimulate innovation if other states succeed with regulatory and related policies that advance economic development.

After the energy crisis in 1973, a number of states established energy policy agencies outside of the PUCs, to promote conservation, new sources of energy, new means of transportation, and to coordinate state government usage of energy. The role of these agencies is similar to new state economic development agencies or task forces; to look beyond the more narrow issues of rates in telecommunications regulation. Partly, as with energy, these agencies can take a longer run planning perspective as they are even more insulated, in contrast to the pressure on PUCs to be concerned about the short-run political issues involved in rate making. What may sometimes appear to be uncoordinated policy fragmentation in a state can actually be positive redundancy (Landau, 1969), providing a different perspective on related issues.

CONCLUSIONS

State actors beyond PUC regulators are now very much involved in telecommunication issues. Most are not involved on a daily basis, as the PUCs are, but telecommunications has become a more salient issue, as the economic development aspect and the information superhighway ideas have emphasized. As a result, governors have become involved in many states. Legislators were already involved because of the need for new laws after divestiture. The type of laws that legislators passed were influenced by political pressures in their states, in predictable ways, as we illustrated in our analysis.

Many state PUC regulators are wary of increased activity by politicians. Gail Garfield Schwartz, former New York State regulator, argued against state legislative involvement: "Once they get in there it is very hard to get them out so I look at our mission as being one of trying to keep ahead of any potential legislative regulatory prescriptions. We do need legislative authority to give us the widest scope for our activities" (quoted in Entman, 1988, p. 14). Other state regulators, such as Bruce Hagan of North Dakota, feel that regulators, like legislators, are political actors, and that legislative intervention is not necessarily bad (Teske, 1987).

Much of the negative perception of non-PUC state actors may come from the legislative bill passed in Nebraska in 1986, which involved extreme pressure politics by U S West. Although this law is seriously flawed, even in this case Mueller (1993) found no strong evidence that Nebraskans have suffered grave consequences from it.

A miniature NTIA on the state level to advise policymakers will stimulate further innovations in the states. It can also provide governors and legislators with a separate set of information and expertise, outside of the rate case environment. The PUC can not really do it because economic development often is construed to mean helping businesses, at least in the short run. Hudson (1990) also noted: "PUCs generally don't have well formulated criteria for assessing socio-economic effects of their decisions and policies" (p. 50).

Economic development competition gives states an incentive to imitate good regulatory changes. As the technological environment of telecommunications changes, continued regulatory changes and innovations will be needed. Many changes are blocked by overly short-term political concerns for those ratepayers who could afford to pay the "real" costs of their service. But if some regulatory changes stimulate economic development,[13] then other states will have an incentive to imitate them, when they see the policy work. State economic development units can analyze and pressure for these changes without appearing to favor one of the entrenched telecommunications interest groups.

As Kahn (1990) recently noted: "In telecommunications it [the cumulative process of deregulation] is reinforced by the competition among the states to attract high-tech industry, which subjects them to the technological imperatives of economic growth, in conflict with historic regulatory policies and goals" (p. 21).

As American businesses and residents become more dependent upon telecommunications to solve problems and coordinate activities, state decision making will become even more important. Getting actors outside of the PUCs more involved can lead to greater experimentation by the states, and as with more narrow regulatory policy issues, each state can handle these problems as they see fit for their particular concerns. Perhaps a model of innovative and well-integrated telecommunications policymaking by governors, legislators, PUCs, and economic development agencies that is superior to that presented by Congress, Judge Greene, the FCC, and the NTIA will emerge from the states.

APPENDIX: MEASURES OF DEPENDENT VARIABLES

1. Specificity of Scope of Law

Q1. Are the PUCs allowed to regulate the telecommunications firms?
    Yes = 1   No = 0

---

[13]And economic theory says they should, if they eliminate dead-weight losses and provide incentives for innovation. See Egan and Wenders for this argument in chapter 5.

Q2. How much specificity is written into the telecommunication rate regulation law? Take into account the following: (a) Does the law say which particular actors to regulate and who not to? (b) Does it say how to regulate (c) Does it say under what conditions to regulate?

Not applicable (no regulation law) = 0
Very specific = 1
Somewhat specific = 2
Not specific = 3

Q3. Does the law mandate the PUCs to deregulate rates and market entry of telecommunication firms?

Yes = 0   No = 1

Q4. How much specificity is written into the telecommunication deregulation law? Take into account the following: (a) Does the law say which particular actors to deregulate and who not to? (b) Does it say how to deregulate? (c) Does it say under what conditions to deregulate?

Very specific = 0
Somewhat specific = 1
Not specific = 2
Not applicable (no deregulation law) = 3

Q5. Does the law allow the PUC to consider alternatives to regulation?

Yes = 1   No = 0

Q6. Does the PUC have the power to reregulate deregulated telecommunication firms?

Not applicable (no deregulation law) = 2
Yes = 1   No = 0

2. Regulatory Instruments

Q7. Do the telecommunication firms have to file price schedules with the PUC?

Yes = 1   No = 0

Q8. Do the telecommunications firms require any certification or permit for operation from the PUC?

Yes = 1   No = 0

Q9. Does the law provide the PUC with the right to inspect all documents of telecommunications firms at any time?

Yes = 1   No = 0

Q10. What kind of investigation process is allowed by the law to the PUC?
Not applicable (no regulation law) = 0
Complaint only = 1
Own initiative under given conditions or complaint and own initiative under
     given conditions = 2
Own free initiative or complaints and own free initiative = 3
No mention of any specifics = 4

Q11. What kind of hearing process is allowed by the law to the PUC?
Not applicable (no regulation law) = 0
Mandatory on receiving complaint = 1
Partly mandatory = 2
Entirely discretionary = 3

Q12. Do the telecommunication firms have to file a report with the PUCs?
Yes = 1   No = 0

3. Oversight Provisions

Q13. Is the PUC required to file a report with the legislature?
Yes = 0   No and No regulation law = 1

Q14. Does the law specify what the contents of the report shall be?
Specific = 2
Not specific = 1
No reporting requirement = 0
Not applicable (no regulation law) = 0

Q15. How frequently do the companies have to file reports?
More than one annually = 3
Annually = 2
Biennially = 1
No reporting requirements = 0

Q16. Does the law mandate the schedules filed by the companies to be open for
public inspection?
Open = 1   Closed = 0

Q17. Does the law provide for a public counsel?
Yes = 1   No = 0

Costs and Benefits of State Regulation

5

THE COSTS OF STATE REGULATION

Bruce Egan
Columbia Institute for Tele-Information

John Wenders
University of Idaho

Since the early days of telephone regulation, state and federal agencies have shared regulatory powers and, predictably, their rules have been both complementary and conflicting. The most difficult issues that arise concern jurisdiction over joint and common costs shared in the production of intrastate and interstate services. Other complicated issues include regulation of operating and financial arrangements—a significant amount of cooperation and coordination is required to make the system work. Much of the cooperation and coordination functions occur through voluntary agreements to split regulatory responsibilities, either by direct state and federal negotiations or through NARUC, the ruling body of state regulatory agencies.

Over the years, as Noam and Geller discuss in later chapters, the courts have often had to rule on the proper separation of powers and jurisdiction. These court rulings are not only driven by issues of states' rights and interstate commerce, but also by concerns over public welfare and national security. The main problem today is defining, jurisdictionally, just what constitutes interstate commerce in telecommunications. Local telephone company (LEC) access lines and services often do not cross state boundaries; in fact, pursuant to the AT&T divestiture, those of the divested BOCs are generally not allowed to. Yet, the services and facilities of the LECs are clearly complementary to the provision of interstate service. An important conflict is brewing that promises to get even worse as new technology allows for network signals to follow software-driven "logical" paths through both public and private networks. It is becoming very difficult (and perhaps soon, impossible) to reconcile these "virtual" network services with political or regulatory boundaries between states.

Although most court decisions have upheld the view that LEC network facilities used in conjunction with provision of interstate services are in fact jurisdictionally interstate, affirming the FCC's power to regulate them, some courts (especially the Federal District Court for the Ninth Circuit) have ruled that physically intrastate facilities and services may be under the regulatory control of states and not the FCC. Such court decisions represent constraints on regulatory reform regarding the future split between state and federal powers in the name of economic and technological progress. Although Congress could rule on these jurisdictional issues, that seems unlikely.

Thus, any consideration of the costs of state regulation and policy recommendations to reduce its scope must account for the constraint of the judiciary's interpretation of state's rights in regulating telecommunications.

The broad goals of both state and federal regulation are generally considered to be to promote and protect the "public interest," which is often construed to mean provision of high-quality, reliable utility services at affordable, nondiscriminatory prices. As discussed later, one of the problems with the public interest approach is that it often becomes the "political interest" once it is thrown into the regulatory system.

Historically, economic theory has been used to guide regulatory policy. Specifically, the implicit function of regulatory agencies is to substitute for free market competition. Economic theory suggests that the efficiency of free market competition leads to maximum social welfare, which is seen to be consistent with the notion of public interest, except perhaps when disadvantaged groups of the consuming public (e.g., poor and "captive" customers) are potentially affected adversely. In many cases public interest considerations of equity and politics drive regulatory policy just as much as economic efficiency. The classic trade-off between regulatory efficiency and equity in public telecommunications concerns the popular "Universal Service" doctrine, promoting broad rate subsidies to basic residential telephone service at the expense of economic efficiency.

We have two goals in this chapter: (a) to investigate the relative efficacy of state and federal regulation in both positive and normative aspects, and (b) make recommendations for changing current regulatory structures.

REDUCING THE COSTS OF STATE REGULATION

As the reader will see, our primary conclusion is that although both state and federal regulation will continue to coexist as institutions, we should try to minimize their impact on otherwise competitive market forces in the telecommunications industry and rely more on the general body of business law to govern any market abuses. Beyond this objective, whatever residual regulation is required is probably best carried out under federal jurisdiction. Obviously, for a variety of reasons, state regulation cannot and will not simply disappear; however, it should be aimed toward (de)regulating to promote procompetitive market processes.

Regulation at both state and federal levels traditionally focused on issues of pricing, costing, and profit levels in a protected monopoly environment. In the new competitive era, regulation should be changed to focus on dynamic market processes. In this new environment, the old formulas for jurisdictional separation of costs and demand are obsolete as private bypass networks and new technological developments allow users to circumvent the old system. New regulatory regimes involving incentive schemes such as simple price ceilings should be adopted to replace traditional rate-base regulation. Regulation should concentrate more on policy coordination and monitoring activities of market players and focus more on the encouragement of market processes and synergies rather than designing new rules for cost allocations in a partially regulated market. Policy coordination and monitoring activities of regulators should be of a passive nature, concentrating on data gathering and dissemination of information. Traditional social issues of equity, universally available service, and public network subscriber complaints probably will continue to be the responsibility of state regulators.

The new regulatory focus should be on oversight of public network issues of compatibility, standards, interconnection, reliability, quality of service monitoring, privacy, security, and anticompetitive activity. It is natural that state regulators would resist such a redefining of their role because it appears to diminish their importance. Although this may be true, the role and political power of regulators will not be diminished greatly if they retain some authority for enforcement of public telecommunications policy.

Within this general refocusing of regulatory activity, there seem to be logical initial separations of state and federal regulatory powers and responsibilities based on their relative strengths to monitor, coordinate, and enforce public communication policy. For example, federal agencies will continue to be the primary coordinator for frequency spectrum policies, which are improving greatly from adoption of new market-based policies. Federal agencies will also be primarily responsible for the standards-setting process of private firms, including those for interconnection arrangements between all private and common-carrier networks.[1] The public telecommunications market is national in scope and coordination of standards is essential. Issues concerning depreciation and cost allocation should simply be eliminated from regulatory jurisdiction altogether as they encourage political mischief and probably worsen social welfare.

---

[1]We do not recommend that government authorities actually get involved in determining appropriate industry standards, only that they devise rules for the private industry to follow that maximize technical progress and innovation and minimize political maneuvering of private interests.

States will continue to be responsible for coordinating their own infrastructure policies, such as promoting interconnection and technology adoption, as well as traditional quality and universal service issues, and any residual franchise and carrier of last resort obligations for local service providers; still, we strongly believe they should play a minimal role even in these areas. In any event, states should eliminate local franchising authority, which constitutes more of a barrier to entry than protection from "wasteful" competition.[2]

Pricing and profits should be gradually deregulated at both state and federal levels and left largely to market discipline from competition. Regarding new and enhanced service providers, neither the states nor federal regulators should be responsible for them; rather, the antitrust laws should be relied on, along with the monitoring and reporting function of regulators and their ability to bring suit or otherwise initiate litigation through state attorneys general.

Throughout the policy analysis to follow, the costs and benefits of regulation are highlighted with the purpose of contributing to the debate by asking what the relative efficacy of state and federal regulation in achieving the goals of regulation is.

THE DYNAMIC COSTS OF REGULATION GENERALLY

In order to discuss the cost of state regulation of telecommunications, we must set a standard for the measurement of costs. In economics, the concept of costs usually used is opportunity costs, or the cost of benefits forgone. This requires an assessment of the alternatives given up in the marketplace, including a specification of what alternative market structures—state or federal regulation or deregulation—are available.

One standard often used to assess economic costs is to measure the neoclassical welfare loss triangles that result from the mispricing of telecommunication services. This approach is instructive, but requires us to assume that static long-run, equilibrium, marginal-cost pricing is the relevant alternative.

Another way to analyze costs is to view the telecommunication marketplace as being in neoclassical long-run disequilibrium; recognizing that the disequilibrium nature of the market provides incentives for innovation and new products. The disequilibrium view notes that any market, no matter how imperfect from a neoclassical static viewpoint, generates ongoing welfare benefits in the form of realized gains from trade. Measures that focus on reducing neoclassical welfare costs may, in the long run, damage the disequilibrium engine that generates growing gains from trade over time, especially in a rapidly changing telecommunications market.

---

[2]Even though the franchising authority of state PUCs is generally not legally exclusive, it is often effectively so, due to bureaucratic processes restricting entry.

A second way to assess the costs of state telecommunication regulation is to ask what the alternatives are. Is it perfect state regulation? Is it perfect federal (FCC–Congress), regulation? Is it imperfect federal regulation? Is it no regulation at all? Even though we do think it is instructive to look at how perfect regulation might reduce static economic welfare losses, perfect regulation or markets are not viable alternatives. In today's political environment, total deregulation is also impossible.

Imperfect, minimal, federal regulation of national and local telecommunications markets will produce greater welfare benefits than continued state regulation of these markets. Our position is based primarily on the grounds of dynamic welfare, not on the grounds that simple neoclassical static welfare losses will necessarily be less than under state regulation.[3] Post-WWII evidence has shown federal regulators to be less obstructive of telecommunication markets than state regulators. State regulators have always been very protective of their control for the simple reason that they, perhaps reflecting the fears of the regulated LECs, are frightened of anything that threatens existing perceived subsidy flows, even if everyone may be better off in the long term.

Our judgment is not based primarily on the idea that federal regulation is likely to deal efficiently with the huge, static, neoclassical welfare losses created by the toll revenue separations process currently used to provide toll-to-local service subsidies. The real benefits from an increased relative role for federal regulation will come from reducing the dynamic costs imposed by protective state regulation that is damaging the growth in gains from trade in these markets.

GOALS OF REGULATION

Regulation, whether by state or federal authorities, should try to emulate a competitive outcome, asking two critical questions. First, what is a competitive outcome and how can the regulators find it? Second, once regulators have been given the power to regulate, how do we prevent politics from replacing the competitive goal with other self-interested goals? The fact is that the telecommunication industry is presently regulated by both state and federal authorities, and if changes are to be made, for better or worse, they must be made through existing institutions, governed by existing laws and political forces.

We first outline how competitive marketplace regulation should emulate and then analyze the real-world political marketplace in which regulation operates. We believe that these perspectives provide valuable guidance.

The usual neoclassical competitive rule for regulation is to set price equal to marginal cost (appropriately adjusted for externalities), and then adjust to meet

---

[3]Although the direct administrative costs of regulation, together with the costs of rent-seeking and defending, are certainly higher with the effects of state regulation added in.

legal revenue requirements using optimal pricing rules (e.g., the inverse elasticity or "Ramsey" pricing rule). The regulatory rule that emerges from this approach is almost wholly a pricing standard. In contrast, the most important thing the competitive standard has to offer for regulation is not a set of pricing rules, but a set of rules by which the competitive process discovers, interprets, and conveys information, which propels the market and creates gains from trade.

The rationale for the pricing rule stems from a Marshallian view of the competitive process. We detail this process, not because we advocate regulators emulating it, but for the opposite reason; we think that, in contrast to Gabel's argument in chapter 2, it is virtually impossible to do so. The reason is that the information requirements of the competitive process are unknown—indeed, unknowable—to an entity such as a PUC. The competitive process should not be viewed as a process for producing marginal cost pricing, but as a process that generates gains from trade, through time, by producing information and providing a mechanism by which market players can act on the basis of such information.

When prices are above the long-run supply price, new resources will earn abnormally high rewards, and therefore resources will be attracted to the market in question. Further, no matter what the state of the existing (observed) market, the lure of potentially high rewards will always induce suppliers to try new products and means of production. Therefore, the possibility of above-cost pricing both provides a lure for investment and innovation, and encourages competitive entry that may prevent it from lasting.

Conversely, when price is below the long-run supply price, the market will be unable to attract or hold productive resources. As resources become more valuable elsewhere or wear out, they will leave the market and not be replaced. This decline in productive capacity will ultimately force prices to rise toward long-run marginal cost.

This description submerges a good bit of useful economics. As stated, the process is timeless. How long it takes for investment, disinvestment, or innovation is simply not specified by the theory—nor can it be. However, some resources are specialized and durable in the short run, and entry and exit cannot be instantaneous. Thus, in trying to apply theory to reality we must understand that markets will usually, and quite naturally, be in long-run disequilibrium in terms of the model being applied.

THE FALLACY OF COST-BASED PRICING

In the process of allocating resources by encouraging entry and exit, competition is said to promote neoclassical economic efficiency by improving price–cost relationships. Maximum economic efficiency results when voluntary exchanges are maximized in any market, each of which leave both parties better off. Economic efficiency results when prices are cost-based, which is why regulators frequently want to set prices equal to marginal cost.

When economists use the term *cost-based pricing* in connection with public utilities, they usually mean long-run marginal cost. Because markets are almost always in competitive disequilibrium in the short run—even those that everyone agrees are reasonably competitive—prices may be either higher or lower than long-run marginal cost. Is it possible to reconcile this disequilibrium aspect of the competitive process with neoclassical economic efficiency? Only if one views the disequilibrium situation as a transitory aberration in the competitive process. First, it must be assumed that the positive or negative rewards associated with disequilibrium will necessarily induce corrective resource flows that always bring the market back into long-run equilibrium. Second, the long-run equilibrium state of affairs must be viewed as the normal state of affairs. In fact, markets are always in disequilibrium.

Thus, the idea of competition as a promoter of efficient cost-based pricing really is not accurate. The idea of pricing at cost, even if the relevant costs were discoverable independent of the competitive process, loses sight of both the dynamic forces that propel the market in the right direction when prices do depart from costs and the incentive that the possibility of above-cost pricing provides for entrepreneurial activity.

At any point in time, a competitive market will contain a variety of firms. Some will do well at fulfilling market demands at a cost level that leaves them with high profits. Others will not do well, and they will have an incentive to imitate and outcompete successful competitors to survive. At any time each firm may have some advantages, which is one desirable feature of competition. This diversity of talent and luck is submerged by the long-run price-should-equal-cost equilibrium view of the market, even though such a long-run view may adequately describe the direction in which the market is going. This is why the long-run, equilibrium view of the firm cannot be used to judge the short-run performance of a market that, by definition, is almost always in long-run disequilibrium.

The main point is that competition is a market process that reveals in a general way what market long-run marginal costs are. In fact, it is probably impossible to discover such costs ex ante; they may be only a theoretical reasoning point to explain what we observe and therefore may not be embedded in any data before us. Such costs are the result of the competitive process, not an input into it, and therefore are not discovered by looking at the firms in the industry at any point in time. It is useful to define market long-run marginal costs as the level to which the competitive process would drive price in the long run if a long-run equilibrium were ever achieved.

This discussion forces the question of where regulators will get the information necessary to regulate either competitive prices or processes in this industry. Unless they are omniscient, they cannot. Information is so decentralized, uncertain, and transitory, that it can only be revealed, if at all, ex post by market processes. In addition, such information may be irrelevant both to the ex ante choices that were in fact made, or to choices made in the future.

Another related point is important in telecommunications regulation. The set of resources and institutions necessary in a truly unregulated market is quite different from those relevant to a regulated market, even if price were equal to long-run marginal cost in each. A competitive market requires firms to prepare for the problems and opportunities encountered by disequilibrium. This implies different human and physical capital and institutions than would be required in a regulated environment, even if such regulation were perfect in the price = marginal cost sense.

Regulation often stifles the development of the capital and institutions necessary in markets perpetually in disequilibrium by not only suppressing the normal competitive alternatives but also the mechanisms by which alternatives develop, appear, and disappear. When a market is deregulated, even if long-run marginal-cost pricing has been the basis for past regulation, it invariably will be left in a state of competitive disequilibrium. Some temporary, neoclassical, welfare distortions may appear but should not be used as an excuse for continuing regulation because this short-run situation will be resolved by competitive forces. Competition is a process that operates from within, where the information and incentives are. It cannot be directed or controlled from without simply because no one can really determine what is going on (much less predict the future).

In summary, if regulators are to regulate according to a competitive standard, they should start emulating the disequilibrium process of discovery that is the heart of the competitive marketplace and economic welfare. This usually prompts economists to suggest that regulators should just deregulate entirely, especially in the fast-moving telecommunications industry. From a theoretical perspective, this is appropriate, but from a practical standpoint, it is not likely. Regulators *can* refocus from a historical, cost-based pricing perspective to one that emphasizes facilitating competitive processes. This would necessarily focus on allowing freedom of entry and exit, freedom of innovation, and the freedom to sell and resell services.

POLITICS IN TELECOMMUNICATIONS REGULATION

Prescribing regulatory reform does not assure that it will happen, in the political process of regulation. In the history of U.S. telecommunications, political competition altered the results of purely economic competition to produce a very politically popular toll-to-residence subsidy. The reason the subsidy scheme was, and is, so popular is that it taxes toll usage, where such usage is heavily concentrated among few subscribers, many of whom are businesses, and uses the proceeds to subsidize the many, politically potent, residential subscribers.

The median toll user (and voter) that drives the political process has far less toll usage than the mean toll user. The subsidy scheme was made possible by: (a) the fact that local residence subscribers are very aware of, and politically sensitive to, their local telephone bills; (b) rapid technological change in provision of toll calling allowed a large subsidy to be generated in toll markets at the same time real toll prices actually fell; and (c) AT&T had little incentive to object to a plan that took revenues from its toll market, and put them into the local BOCs, effectively moving money from one pocket to another.

In a regulatory balance, firms received relatively safe, guaranteed profits and regulatory protection from competition and antitrust laws, and regulators got a cross-subsidy from the relatively few heavy toll users to the relatively many local residential subscribers. At the same time, the overpricing in toll markets created large efficiency distortions there, whereas underpricing removed competitive pressures in the local market. The protectionist policies of state regulators made local competition for many services simply illegal. Further, with no competitive pressures to match prices with costs, statewide averaging was the easiest way to price most local services, and some local prices—mostly business—ended up being well above the cost of providing local service with the newest technology.

In summary, these pricing developments have probably been politically efficient in getting votes to drive the political-regulatory mechanism. In order for substantive changes to be made, we must look to a change in the balance that produced this politically efficient outcome. It is not enough for economists to make utopian recommendations that presume an ideal political marketplace.

Historically, despite neoclassical welfare and dynamic losses, competition was opposed by nearly all parties: the Bell System, local regulators, Congress, and the FCC. Eventually the FCC, with significant prodding from the courts, succumbed to the market forces of competition. Local regulators did not. They continuously opposed nearly every concession to competition,[4] and the FCC had to preempt numerous attempts by the states to assert their sovereignty in this regard. There have been some notable exceptions. For example, the Illinois Commerce Commission (1992) proposed a model for introducing local network competition called a Telecommunications Free Trade Zone (TFTZ), which, however, is limited to the downtown area of Chicago. Another exception is the relatively aggressive New York State Public Service Commission approach to easy and efficient access for independent private network operators to interconnect with the public switched telephone network.

The FCC today takes a broader view of telecommunications policy, in part because the FCC does not hear the median toll user as loudly as do state regulators. The FCC is more likely to look at all the costs and benefits of various alternatives in performing its inevitable balancing act. In this sense, then, more of an emphasis on federal regulation should be an improvement, *ceteris paribus*.

Still, the FCC oversees a toll market pricing regime that continues to include large static, neoclassical welfare losses. The FCC's attempt to reduce them with larger subscriber line charges has been stalled for the foreseeable future. There-

---

[4] A good example of state resistance to a key federal deregulatory policy is the well-known 1970 "cream-skimming" case, where NARUC filed suit to block the FCC's Specialized Common Carrier Decision allowing carriers like MCI to sell long-distance telecommunications on private networks to third parties.

fore, we do not think that, for example, having the FCC preempt and regulate all toll traffic and carrier access charges would change the current situation much, although interstate toll prices fell much more over the last decade than intrastate toll prices. Our hope is that market activity will continue to threaten to pick the separations and carrier access charge processes apart, upset the political-economic balance at the margin, and cause welfare-improving changes, such as those resulting from FCC decisions in the early 1980s.[5]

There is considerable evidence that state regulation is very costly, primarily because of the exclusive local franchise barriers to entry that persist in almost all jurisdictions. Thus, on the whole, the LECs and their state regulators have thwarted free entry and resale of local services that would prevail in an otherwise competitive environment. In the area of intrastate, interLATA toll, some states have deregulated retail toll services, but not as much as the FCC. And a number of states still flatly prohibit intraLATA toll competition.[6]

Although most states have flatly prevented some innovations, such as privately owned shared tenant services, this approach does not get to the main problem, which is that we simply do not know what states have prevented from happening. Until the market is opened up we cannot know what will emerge; we do not know what niches and latent demand are out there until entrepreneurs are allowed to explore the possibilities. No one could have predicted the latent demand for various kinds of terminal equipment and services that has emerged pursuant to the FCC's Registration program in 1972 or the niche found by value-added services such as the alternative Operator Service Providers (OSPs).

COMPARING FEDERAL AND STATE REGULATION

The rate at which regulators accommodate technological and competitive changes, rather than fighting such change, is an indicator of progressive (de)regulation. It is beyond the scope of this chapter to make a micro-assessment of the ex ante intentions of state and federal regulators in recent years.[7] Rather we examine, ex post, the impacts of recent policy decisions in the postdivestiture period on productivity, prices, profits, service quality, and entry barriers.

---

[5]Such as the FCC's MTS/WATS Market Structure proceeding CC Docket 78-72 (the "access charge" case).

[6]What happens at the retail level becomes largely immaterial with carrier access charges in place to effect such a subsidy. In other words, the fact that there are many competitive alternatives at the retail level does not necessarily lead to an economically efficient outcome. The same situation exists in cellular mobile radio telephone service, which features many retail service vendors, but which is a duopoly at the wholesale level.

[7]There is some recent theoretical literature on modeling the pricing decisions of state regulators that may prove useful for predicting the outcome of simultaneous federal and state regulatory interaction. Garber (1990; Garber & Peterson, 1990) conducted positive theoretical analysis to investigate the incentives and intentions of regulators that give rise to specific policy preferences. This work presents some models of pricing behavior of state regulators and how they are affected by Federal pricing policies.

The literature on productivity trends is decidedly mixed. Crandall (1991) and other researchers find significant productivity gains due to deregulation and competitive entry. Others found little or even negative changes in productivity since the AT&T divestiture.[8] The results of such productivity studies are highly sensitive to the data used, which is often of questionable reliability.

Price trends are easier to observe and evaluate. In Table 5.1 we show a summary of state rate case activity for local telephone companies from 1984–1991. The overall frequency and level of rate increase requests fell substantially after 1984 to almost nominal activity by 1987. This activity increased slightly from 1987 to 1989, and while rising substantially in 1990, was not sustained in 1991. These trends are indicative of the fact that LEC cash flow and net income have been quite high in recent years due to very low inflation, tax reform, certain accounting changes, and increases in depreciation allowances. In fact, as Table 5.1 shows, state regulators actually ordered overall rate reductions in the years 1987–1990.[9]

Even though the average total monthly bill of residential and business telephone subscribers has been stable since 1984,[10] rates have been "rebalanced" with higher local and lower toll rates. The benefits of the lower toll prices have

---

[8]In a paper presented January 24, 1992, at the Columbia Institute for Tele-Information Conference on Private Networks at Columbia University, Greenwald suggested that most productivity study results were suspect and that his own examination of the data showed slower productivity growth in the industry postdivestiture. In a recent regulatory proceeding to determine the viability of competition in Canada (CRTC 92-12), a host of academic researchers presented studies and testimony on U.S. telecommunications productivity in the postentry period (after 1976) and postdivestiture period (after 1983) with mixed results and no firm consensus.

[9]One effect of the convenient coincidence of relatively low inflation and interest rates and unprecedented economic growth, with the industry-related factors of tax and accounting reform, was to stimulate many voluntary price cap regulatory contracts between state and federal regulators and telephone companies. Even though most states will no doubt actively consider regulatory reform in favor of incentive-based regulation, in practice the plans are really just modified versions of traditional rate base regulation. Of the more than 30 states that have implemented incentive plans, the majority of them do not completely eliminate price and profit controls typical of rate-of-return regulation. Rather, they only eliminate it on some discretionary services that represent a minor portion of the business. In fact, all of the plans in effect include periodic rate base calculations and profit reviews, effectively extending the regulatory lag of the old regimes.

It is not clear that the rapid rise in incentive-based regulation—as a voluntary agreement between telephone companies and regulators—would have occurred if inflation and recession prevailed during the postdivestiture period. The healthy cash flow situation of the LECs allowed many incentive regulation schemes to include large up-front reductions in basic subscriber rates, eliminating much of the potential consumer resistance. The staying power of the new regulatory regimes has not yet been tested for extended periods of low telephone utility earnings. There is one observation from recent history. New York Telephone experienced very bad earnings under their first experiment with price cap regulation and the company could not come to an agreement with state regulators to extend the price cap agreement. Nevertheless, the overall trend of states to move toward incentive regulation is still the right direction for improved economic efficiency.

[10]In fact, even the monthly bill of the poor and elderly has remained stable and in many other consumer groups it actually fell (although for median customers it likely rose). See: Larson, Makarewicz, and Monson (1989).

TABLE 5.1
State Telephone Rate Cases (Dollar Amounts Shown in Millions)

Year		Revenue Requests	Increases Granted	
1984		4,023.7	3,875.5	
1985		1,627.2	1,154.9	
1986		643.7	290.0	
1987		146.3	(519.0)	
1988		357.9	(1,366.4)	
1989		447.0	(838.5)	
1990		1,109.2	(451.1)	
1991	First	184.3	2.8	(Pending 372.4)
	Second	2.1	7.9	(Pending 219.2)

Reprinted by permission from FCC (1991).

given rise to large increases in toll calling with no decrease in basic local service subscribers. In fact, basic telephone penetration is up 2.4% since 1984.[11] The obvious result of all of these changes is large gains in economic welfare.

Table 5.2 shows nominal price change data for regulated local and toll telephone services for 1984–1991. Local rates rose about 45% (including the FCC mandated SLC) and interstate toll rates fell by about 39%. Intrastate toll rates fell by about 7% overall. Thus, the states have not rebalanced rates as much as the FCC in its interstate jurisdiction.

Interstate toll minutes of use rose an average 13% per year since 1984, and about 10% for intrastate toll.[12] Even though intrastate toll traffic historically exhibited lower growth rates than interstate traffic, increased competition in services and prices probably caused most of the growth differential. In the interstate market, alternative service options abound, especially in urban areas, as equal access has largely been achieved. In intrastate jurisdictions, the majority of states have allowed toll competition in both interLATA and intraLATA markets, but no state has yet implemented equal access for intraLATA toll services (which represents about one fourth of total switched toll minutes of use).

The postdivestiture competitive period gave rise to many new toll service options for customers, with about 450 toll carriers across the nation. At the same time, the overall quality of telephone service has been maintained or even improved. Overall customer satisfaction is higher and technical performance of the public network has improved when comparing 1989 to 1984 results.[13]

---

[11]Of course, there were many "lifeline" discount rate plans implemented during this period.

[12]Data on growth rates for toll usage may be found in the FCC's Semiannual Report on Telephone Trends (1991).

[13]See the quality of service trends in the FCC report *Update on Quality of Service for the Bell Operating Companies* (1990). Basic service installation delays, however, are up slightly over 1984 levels.

TABLE 5.2
Telephone Prices (Annual Rate of Change)

Year	Local	Intrastate Toll	Interstate Toll	Access*
1984	+17.2	+3.6	−4.3	—
1985	+8.9	+0.6	−3.7	−8.1
1986	+7.1	+0.3	−9.5	−14.3
1987	+3.3	−3.0	−12.4	−21.7
1988	+4.5	−4.2	−4.2	−8.5
1989	+0.6	−2.6	−1.3	−14.2
1990	+1.0	−2.2	−3.7	−3.8
1991	+2.9	−2.3	−2.2	−4.2
Total	+45.5	−7.5	−39.1	−74.8

*Interstate only.
Reprinted by permission from FCC Monitoring Report (1991). CC Docket No. 87-339.

We can estimate static welfare gains from regulated rate rebalancing between 1984 and 1991. During this period, the FCC implemented an access charge plan in CC Docket 78-72. This plan phased in subscriber line charges (SLCs), effectively increasing residence and business local monthly charges, and phased out a portion of the subsidy from carrier access charges for interstate toll service—and in turn—the retail interstate toll rates of interexchange carriers. The exact (nominal) price changes that occurred in the rate rebalancing program appear in Table 5.3.

LEC switched access rates for toll carriers fell about 59% after 1984, from almost $.17 per minute of toll usage to about $.07 by 1992 (average of sum of total per-minute switched access charges for both originating and terminating ends). At the same time, AT&T retail residential toll rates fell more than 39% for interstate service. Carrier access charges represent some 30–50% of retail toll charges per minute of use. As such, the retail toll price reductions, in absolute terms, actually fell about the same as LEC switched access charges, which have fallen by almost 75%. Residential SLC went from zero to $3.55 per line and business SLC went from zero to $3.57 from 1984 to 1990.

Interestingly, during the same period *intrastate* interLATA switched access charges were about the same as interstate, and yet retail intrastate toll charges fell only nominally. The reason many state jurisdictions set intrastate access tariff rate levels at (or near) parity with interstate levels is concern over customers' arbitrage of the difference between the two rates. Due to operational problems, it is often difficult for LECs to distinguish interstate from intrastate toll minutes of use. One plausible explanation of why intrastate access tariff price reductions are not fully reflected in reduced intrastate toll retail tariffs is *intraLATA* toll prices (as distinguished from intrastate *interLATA*) are maintained at relatively high levels. Accurate price data on average *intraLATA* toll prices for the United States are not available.

TABLE 5.3a
Interstate Charges by Local Telephone Companies to Long-Distance Carriers
(National Average for "Premium" Service in Cents per Minute)

Dates	Carrier Common Line Charges Per Originating Access Minute	Carrier Common Line Charges Per Terminating Access Minute	Total Traffic Sensitive Charges Per Access Minute	Total Charges Per Conversation Access Minute
05/84 to 12/84	5.24	5.24	3.1	17.3
01/85 to 05/85	5.43	5.43	3.1	17.7
06/85 to 09/85	4.71	4.71	3.1	16.2
10/85 to 05/86	4.33	4.33	3.1	15.4
06/86 to 12/86	3.04	4.33	3.1	14.0
01/87 to 06/87	1.55	4.33	3.1	12.4
07/87 to 12/87	0.69	4.33	3.1	11.5
01/88 to 11/88	0.00	4.14	3.1	10.6
12/88 to 02/89	0.00	3.39	3.1	9.8
02/89 to 03/89	0.00	3.25	3.0	9.6
04/89 to 12/89	0.00	1.83	3.0	9.1
01/90 to 06/90	1.00	1.53	2.5	7.8
07/90 to 12/90	1.00	1.23	2.5	7.5
01/91 to 06/91	1.00	1.14	2.4	7.2
07/91 to 12/91	0.96	0.96	2.4	7.0

Reprinted by permission from FCC Semiannual Study of Telephone Trends; FCC Monitoring Report (1991), CC Docket No. 87-339.

TABLE 5.3b
Interstate Subscriber Line Charges (by Local Telephone Companies
to End Users, in $ Per Month Per Line)

Dates	Residential and Small Business	Multiline Business	Centrex
05/84 to 06/84	0.00	4.99	2.00
06/85 to 09/85	1.00	4.99	2.00
10/85 to 05/86	1.00	4.97	2.00
06/86 to 12/86	2.00	4.97	3.00
01/87 to 06/87	2.00	5.12	3.00
07/87 to 11/88	2.60	5.12	4.00
12/88 to 03/89	3.20	5.12	5.00
04/89 to 12/89	3.50	4.94	6.00
01/90 to 06/90	3.48	4.84	6.00
07/90 to 12/90	3.48	4.83	6.00
01/91 to 06/91	3.48	4.77	6.00
07/91 to 12/91	3.49	4.74	6.00

Reprinted by permission from FCC Semiannual Study of Telephone Trends; FCC Monitoring Report (1991), CC Docket No. 87-339.

TABLE 5.3c
Average Monthly Residential Telephone Rates

Charge	1983	1984	1985	1986	1987	1988	1989	1990
Unlimited Local Calling	10.50	12.10	12.17	12.58	12.44	12.32	12.30	12.40
Subscriber Line Charge	00.00	00.00	1.01	2.04	2.66	2.67	3.53	3.55
Taxes	1.08	1.25	1.36	1.51	1.56	1.58	1.70	1.83
Total	11.58	13.35	14.54	16.13	16.66	16.57	17.53	17.78
Lowest Available Monthly Rate (includes SLC and taxes)	5.93	6.20	7.46	8.84	9.41	9.25	10.23	10.35

Reprinted by permission from FCC Semiannual Study of Telephone Trends; FCC Monitoring Report (1991), CC Docket No. 87-339.

Although it seems impressive that interstate switched access charges fell by 75% after divestiture, in fact the charges are still quite high in absolute terms (about $.07/min). After 1989, political resistance against further SLC increases on local service has prevented the FCC from continuing this rate rebalancing.

Prior to divestiture, average intrastate toll prices were somewhat lower than the average price for interstate toll. Due to the rebalancing of rates between 1984 and 1991, interstate prices are now much lower than intrastate, even though the average distance of the call (and presumably the costs) is less in the case of intrastate. This shows how inconsistent regulatory policies between federal and state jurisdictions can seriously distort price–cost relationships across markets.

We present the data required to calculate the static welfare gains from rate rebalancing between 1984 and 1991 in Table 5.4.[14] Our straightforward analysis of the data yields the welfare gains and losses presented in Table 5.5 for each market segment for real price changes since 1984. The sum of these changes in welfare is about $11 billion. The combination of toll tariff rate reductions and inflation caused very large real price decreases and demand stimulation, all of which increase the welfare gains. Inflation since 1984 has nearly offset the nominal increases in business tariff rates for basic local service. Residential local

---

[14]Estimated price elasticities (E) for broad market segments are presented, along with nominal and real price (P) and quantity (Q) data. Prices given for local service are average monthly rates for basic service. Toll prices are average charges per minute of toll usage. Quantity data for toll minutes is for the year 1991, whereas for local service it is lines in service at the end of the year. The price changes (dP) are in real terms (the Consumer Price Index rose by 33% since the end of 1983). The marginal cost estimates are taken from prior studies and represent rough estimates of average incremental costs. Even if there is a great deal of error in the marginal cost estimates, the majority of changes in welfare are due to the very large differences in price elasticities, in which we have much more confidence.

TABLE 5.4
Input Data (1991)

	Local		Toll	
	Business	*Residence*	*Intrastate*	*Interstate*
	$41.09/line/mon	$17.78/line/mon	$.23/min	$.20/min
	$30.89	$13.37	$.17	$.15
	40M lines	100M lines	125B mins	147B mins
1983–1990 price change	+8.5%	+15.4%	−40.5%	−72.1%
Elasticity	−.10	−.04	−.6	−.8
Marginal Cost	$25.00/line/mon	$25.00/line/mon	$0.03/min	$0.04/min

service price increases have little effect on total welfare changes because of the very small price elasticity.[15]

The welfare gains in Table 5.5 are high compared to prior studies that showed welfare losses under predivestiture price structures to be in the range of $2–10 billion.[16] The reason is that substantial toll market growth since 1990 increased the total level of social welfare.[17] This rough analysis illustrates the potential size of static welfare gains from rate rebalancing. Although these static gains are substantial, recall that dynamic gains from competitive markets that are in constant disequilibrium may yield much greater total welfare gains.

Returning to issues apart from rates, regulatory concerns that deregulation, competition, and price caps would cause levels of infrastructure investment, rates of technology adoption, and perhaps quality of service to fall dramatically, are not only unfounded, but, based on the experience of AT&T after 1983, the opposite of what actually happened. Infrastructure investment in public networks by LECs may in fact rise, and wasteful spending and expenses would fall, in response to the competitive threat from deregulation and open entry.

---

[15]There are other factors that make these welfare loss estimates conservative. Only tariff rate reductions are examined here. Presumably competitive entrants, many of which do not even file tariff rates, have lowered the effective market prices even more as they captured significant toll market share from AT&T and the BOCs over the study period 1984–1991. In addition, on the cost side of the calculation, technology has been advancing to decrease real unit costs for toll and local service, thereby lowering the assumed cost curve and increasing welfare gains even further.

[16]There are many different studies with markedly different assumptions leading to a wide range of welfare loss estimates. See, for example, Wenders and Egan (1986), Rohlfs (1979), and Griffin (1982).

[17]If the calculations were performed for the demand curve that existed prior to divestiture, the welfare gains would be cut by almost two thirds to about $3.7 billion.

TABLE 5.5
Estimated Changes in Social Welfare, 1991 (1983 price levels to 1991 price
levels adjusted for inflation)

	Local		Toll	
	Business	*Residence*	*Intrastate*	*Interstate*
Change in welfare	$-37 million	$100 million	$4.5 billion	$6.5 billion

JURISDICTIONAL ISSUES

We have argued that minimal regulation is the preferred approach at all levels, but that federal regulators are more likely to pursue that policy. But, in the absence of explicit Congressional legislation, the courts are establishing the regulatory jurisdictions. In our view, they are moving in the wrong direction.

A 1990 court decision about enhanced services is especially problematic. The Federal District Court for the Ninth Circuit ruled that the FCC Computer Inquiry III decision was not appropriate and that the FCC cannot preempt state authority to regulate enhanced service providers. This decision implies that the FCC also cannot preempt the states on ONA and other policies. Absent the kind of coordination Noam and Geller call for in later chapters, this could result in a hodgepodge of state regulations that would adversely affect the national marketing plans of firms in the burgeoning enhanced services industry. It is alarming to think that national telecommunications policy may be based on a patchwork of case-by-case court decisions.

Uncertainty about market entry and restrictions increase risk for firms considering large capital expenditures. Competitors fear that if the giant LECs are allowed to enter new markets, they could get squashed. In turn, the LECs, spending their political energies on convincing policymakers to lift operating restrictions on entry into high-growth toll and information service markets, are slow to offer efficient interconnection and distribution facilities for the use of others who represent potential future competitors.

In the absence of a clear competitive policy, present regulations and court decisions provide a forum for rent-seeking and rent-defending activities as the various players jockey for advantage. Every time policymakers hand down a key policy decision involving adversarial parties, no matter how well intentioned, winners and losers are created. The winner enjoys a market advantage and the loser is crippled in its ability to compete. Over time, the winners become an entrenched interest group that fights hard to maintain their competitive advantage under the decision. This makes it difficult for policymakers to consider changing the original decision in response to changed market or technological circumstances.[18] The more regulators, the worse the problem becomes.

CONCLUSIONS

We recognize that complete deregulation is not feasible, and may not be completely desirable, but the best form of regulation would attempt to let the market operate wherever possible. Policymakers should pursue other policy goals in the least obtrusive fashion so that the dynamic gains of competition can be achieved. Federal regulators seem most amenable to this approach, and, therefore, we would like to see them making more of the critical decisions, and state regulators doing less. Court decisions moving in the opposite direction are misguided and likely to cause harm to consumers and society. Congress and the president can lead on this critical issue, but must be willing to fight difficult political battles they have not addressed successfully in the recent past.

In particular, we believe policymakers should: (a) remove telephone company line of business restrictions on toll, manufacturing, video, and information services; (b) remove exclusive franchises for cable television or telephone companies to promote competition; (c) abolish rules prohibiting financial and operating arrangements between cable and telephone companies, so that cooperation can occur (analogous to the Japanese model of "cooperative competition," where firms may cooperate but will be subject to direct competition from other such cooperatives); (d) develop and enforce rules for efficient nondiscriminatory interconnection and resale between competitive networks and public telephone networks; (e) implement price cap regulation to eliminate incentives for cross-subsidies among regulated and unregulated lines of business; and (f) play a more active role in helping firms to adopt appropriate technological standards for public network providers.

---

[18]The landmark ENFIA decision made two decades ago is a case in point. To "protect" new competitive long-distance carriers, the FCC ENFIA decision established a 55% discount on LEC access charges for all firms except AT&T to give them a competitive advantage at the margin. As the "winner" in the ENFIA decision, the competitive toll carriers fought hard to prevent the gradual phasing out of the ENFIA discount.

Similar results have occurred in many other key decisions. The Cable Television Act of 1984 is another example of an original decision designed to protect a fledgling industry that "grew up" and became a strong monopoly power with enough lobbying clout to keep the protectionist legislation in place until Congress lost patience with steady rate increases higher than the cost of living, and revoked much of the 1984 legislation with a reregulatory 1992 law.

6

THE BENEFITS OF STATE REGULATION

Sharon B. Megdal
MegEcon Consulting Group

Most agree that predivestiture regulatory practices are not appropriate for the postdivestiture U.S. telecommunications industry. There is disagreement, however, as to how to effect the necessary modification of regulatory procedure. This disagreement has been evident among economists and regulators alike. Underlying the debate has been the structure of regulation established by the Communications Act of 1934. State authorities are charged with regulating intrastate communications, and federal authorities regulate interstate communications, with some preemptive powers.

The courts, charged with upholding existing laws, have influenced the debate significantly. Judicial actions have not only determined industry structure, they have limited both state and federal regulatory authority. Telecommunications regulation is accomplished through a marbled structure. Federal and state actions intertwine to determine (at least in part) telecommunications offerings, rate schedules, and profits. The current system of shared regulation is criticized for its cost, complexity, inertia, and contentiousness. A key question is: What is the appropriate regulatory model? Is it a fully deregulated environment or a partially regulated environment? If the industry is to be partially regulated, the likely scenario for some segments of the industry for the next several years, who should do the regulating? Do the states still have a useful role to play? This chapter focuses on the last question, assuming a fully deregulated environment is still several years away. It is also assumed that the system of regulation will involve federal regulation; that is, a system of only state regulation is not an option.

A RATIONALE FOR A MARBLED STRUCTURE

There are three elements of the role of government in a market economy—stabilization, income distribution, and resource allocation. Analysis of the stabilization function of government is central to macroeconomic policy. Discussion of the proper role for government in (re)distributing income is inherently normative. This role for government has some relevance to regulatory policy designed to promote universal service, but it is not be a focal point of this chapter. Examination of the role for government in resource allocation, on the other hand, involves significant positive economic analysis and has direct bearing on the issue of the appropriate jurisdiction for telecommunications regulation.

There are three fundamental situations where government intervention may correct for the failure of private markets to allocate resources efficiently: public goods, externalities, and natural monopoly. The means for correcting these problems are the subject of much scrutiny. Relevant to the topic in hand is the potential for government regulation to correct for market failure due to natural monopoly.

The economic theory of fiscal federalism, perhaps best espoused by Oates (1972), further argues that centralized government is most appropriate for carrying out the stabilization and income redistribution functions. The best level of government to correct for market failure, it is argued, depends on the nature of the situation. For example, when dealing with a national public good, such as defense, it is most appropriate for the federal government to intervene. In dealing with more local public goods and externalities, the efficiency gains from a less centralized approach may outweigh any cost savings associated with central government involvement. If, however, tastes are uniform across jurisdictions, a centralized approach may be appropriate. A uniform or centralized approach, however, will not take into account geographic diversity in tastes or technology. The policies of subnational governments can incorporate diversity. Moreover, local government activity will result in competition among jurisdictions and more policy innovation.

As noted by Representative Edward Markey (1990), "Congress must set a consistent, timely, and comprehensive national telecommunications policy, a policy that ensures the principles of universal service, diversity, and localism—the cornerstones of the Communications Act and the foundation on which the world's greatest telecommunications network was built (p. 1).

When regulation of telecommunications involved setting prices and profits for providers of "plain old telephone service," most Americans received service provided by AT&T. The network was truly a national network, and rates were determined by both federal and state regulators. In the current environment, the network is still effectively a national network in that it is seamless from the user's perspective, but there are multiple owners of the network and there may be parallel networks. The long-distance market is characterized by competing companies providing parallel networks, with a commonality being their connection to the local network. The local network is largely characterized by monopoly ownership of nonduplicated facilities that serve millions of customers over several jurisdictions. Rates are still set by federal and state regulators, often operating according to laws that were established in the old environment.

The ownership and structure of the U.S. telecommunications network are unique. The current industry structure presents challenges to those charged with developing appropriate regulatory policy. There can be no doubt that telecommunications services and infrastructure are vital to our nation's prosperity and competitiveness. Does the marbled system previously described no longer serve the public well? Is federal regulation superior to the marbled structure of regulation? Is state regulation an anachronism?

An evaluation of the alternative models for regulation must be conducted in the context of policy goals. Although establishing the goals may involve controversy, let me suggest four goals against which regulatory policy can be gauged:[1]

1. Provision of appropriate customer safeguards.
2. Incentives for efficiency and innovation.
3. Facilitation of competition.
4. Reasonable administrative burdens and costs.

The first goal, provision of appropriate customer safeguards, is a very broad one. When access to the local network is provided under largely monopoly conditions, it requires that there be safeguards for retail customers and for competitors who may rely on local exchange company (LEC) services in order to provide service or compete with the LEC. The protections may relate to prices, service quality, conditions of service, and complaint oversight. The second goal recognizes that it is important for the regulatory environment to encourage or at least not inhibit technological efficiency and innovation. The third goal acknowledges the importance of the competitive process and the necessity to facilitate or at least not impede the emergence and development of competition. Although competition may not always be an end in and of itself, the competitive process should allocate resources wherever feasible. Though regulatory costs may be difficult to gauge precisely, they are significant. It would be desirable to reduce them, without substantially sacrificing other regulatory goals.

At a theoretical level, let us consider a regulatory model where all telecommunications regulation would be carried out at the federal level. Regulation of interstate services at the federal level would continue as before, but now regulatory policy for intrastate services would be set nationally. Certainly all the benefits of the current system of federal regulation would continue. Quality and interconnection standards could be set uniformly for the nation. There would be some noticeable changes in the regulation of intrastate services, as the stringency of regulation would be uniform. Service offerings would not depend on individual

---

[1]See Megdal and Lain (1988) for a more complete discussion of these goals in the context of state regulatory reform initiatives for local exchange companies.

state regulatory policies. For example, intraLATA competition would either be allowed everywhere or prohibited everywhere. The availability of a service, such as AT&T's Megacom service, would not be limited by a single state's refusal to approve it.

Because telecommunications services are provided by private businesses, the central regulator could not impose uniform national prices for local service. The federal commission, if it chose to regulate prices at all, could continue the practice of establishing rates (or limits on rates) for each state. Perhaps there would be an effort to impose uniform prices for an entire regional holding company or interexchange carrier. This approach would be less costly to administer but, to the extent costs vary across the states served by a single company, would result in greater deviation of price from cost. At issue, then, is the institutional structure for policy determination and oversight. One option is a massive federal bureaucracy, located in Washington, that would handle all policy matters. An alternative is the creation of state or regional field offices that would conduct the business of regulation, following centrally determined policies and procedures. The administrative costs would also depend on the hearing practices of the federal regulators. A common administrative practice at the state level is the hearing process. This involves both public hearings, where the public can participate, and administrative law proceedings. If the centralized regulatory authority were to dispense with such hearings on state-level matters, administrative costs would be reduced considerably, at the expense of the ability of some to participate in the process. Would the administrative costs associated with the centralized approach be any lower than with the current decentralized approach? They could be, but it is not clear that they would be. Although it may be tempting to conclude that one bureaucracy is cheaper than 50 bureaucracies, federal rules and requirements can be very costly to administer. In addition, there are economies of scale to state regulatory commissions, as they regulate more than one industry. Therefore, the relative administrative costs are not clear.

It would be expected that centralized regulation would effect a uniform approach to product introduction, pricing, profit oversight, and competition. This uniformity has advantages and disadvantages, which depend on the appropriateness of the federal policy and the variability in conditions across the nation. For example, the federal policy could be inappropriate for current market conditions and allow a monopoly provider of essential services to price without constraint. This could be the case nationally or on a more regional level. Federal regulators could, on the other hand, continue strict price regulation for services subject to competition and impede the competitive process for the nation as a whole. A uniform regulatory policy that ignores the incentive problems associated with traditional rate-of-return regulation could likewise be problematic. Moreover, federal policy could be subject to variability in approach, depending on the policy goals of the current administration. Although some actions cannot be undone, it is clear that reregulation of deregulated entities or services is not out of the question. Certainly, a uniform national policy that is appropriate to market and industry conditions nationally would be superior to a fragmented approach that does not provide proper incentives for efficiency and innovation or does not facilitate competition sufficiently. Again we see that, in terms of the first three goals, federal regulation is not necessarily superior and may be inferior to the current marbled system of regulation.

Because we have not operated with only federal regulation of the telecommunications industry, we do not know how much flexibility Congress would extend federal regulators under this alternative approach. The recent experiences of the savings and loan and cable television industries suggest that Congress will attempt to intervene when regulatory problems are perceived. Associated with the creation of the Resolution Trust Corporation was a massive bureaucracy. Cable television regulation has been reintroduced. Replacement of state regulation with federal regulation involves some significant uncertainty as to the autonomy the federal regulatory authority would be afforded.

Theoretically, therefore, it is not clear whether the centralized approach to regulation is superior to the current marbled system in terms of the four goals enumerated earlier. Finally, a centralized approach would not provide the laboratory that 50 states do.

THE STATES AS LABORATORIES

If theory is unable to provide us with clear guidance regarding the preferred approach to regulation, what can experience under the marbled structure tell us? The current system of regulation is an imperfect and evolving process. It has, however, revealed much to us. Although some would complain about the pace of change, state regulators have responded to the call for regulatory reform. In the 1980s, while the FCC was talking about deregulation, states were doing it. States were performing their laboratory function. Several states, observing a more competitive long-distance market, quickly lessened restraints on AT&T. Alternative approaches to regulating local exchange carriers were advocated, debated, modified, and adopted.

For example, in NYNEX territory, the New York approach to incentive regulation was implemented simultaneously with the social contract approach of Vermont. In response to the perceived needs of each state, two very different regulatory methods were adopted for units of the same company. The Vermont approach required statutory change and represented a more radical departure from traditional practices than did the New York policy. For a 3- or 4-year period, the Vermont policy eliminated rate-of-return regulation, capped basic service prices, allowed pricing flexibility for competitive and nonessential services, and required digital switching throughout the state. By eliminating profit oversight entirely, Vermont offered New England Telephone significant incentives to operate efficiently and market effectively. Pricing flexibility gave the company the tools by which to compete and market its products. The investment program recognized service quality issues, and through the cap on basic rates and other provisions, customer protections continued. New York continued rate-of-return regulation, but allowed for some upward flexibility in earnings. At the same time, New York Telephone promised not to seek any general rate increases for a set period of time. The New York approach, by allowing the company to keep earnings above the traditional level, provided incentives for efficiency and innovation. Unfortunately, by not incorporating any pricing flexibility, the plan limited the tools the company could employ to improve its earnings. Both approaches served as models for other states.

In U S West territory, different policies were advocated. A push toward service-by-service deregulation and pricing flexibility was made at the same time one U S West state, Nebraska, enacted far-reaching deregulation legislation. The service-by-service approach relaxes regulation for particular services. Services might be deregulated, where the service's price is no longer regulated (except perhaps for a requirement that price cover incremental cost) and the service's costs are separated from the regulated rate base. Alternatively, pricing flexibility would be accomplished through detariffing, where prices could be changed with minimal notice, but the service's costs are not separated from the regulated rate base, and revenues from that service are included in the company's allowed revenue requirement. Other pricing provisions include flexibility within established bands or discounts from tariffed rates. The Nebraska approach was unique in that it removed nearly all commission authority to review profits and rates for the state's largest local exchange carrier.

Other regional holding companies were not as quick to introduce proposals, but when they did, they and their state commissions often were innovative. For example, California became the first state to adopt an index approach for pricing basic local service, similar to the formula adopted by the FCC for AT&T. The California plan continued profit oversight, but included provisions for the sharing of earnings above an authorized level. Kansas, like Vermont, abandoned rate-of-return regulation for the duration of an alternative regulation trial. In Kansas, basic service prices were frozen for a 5-year period, and pricing flexibility was allowed for a limited set of services. Network investment was also required. Other states have, for an established period of time, combined pricing flexibility for competitive or nonessential services and earnings sharing with a freeze on the local service prices. Changes in the approach to state regulation of long-distance carriers have evolved simultaneously.

Clearly, the major challenge for state regulators over the next decade is continuing appropriate reform of local exchange carrier regulation. States are reviewing and revising their alternative regulation plans. Legislatures are changing the statutory rules for regulation. In response to legislative changes and an application from New Jersey Bell, New Jersey regulators replaced the 1987 Rate Stability Plan with a much more comprehensive, longer term plan that combines significant protections, deregulation of competitive services, and a long-term program of infrastucture investments.

Data from different types of regulatory regimes will be available for analysis. A uniform federal policy would not provide such variation in data. For example, the sharing provisions included in state incentive plans vary considerably from state to state. In some states, the sharing is based on the rate of return on equity. In others, it is based on the rate of return on investment. Some sharing is closed-ended, meaning all earnings above an established rate of return are returned to customers. Other sharing is open-ended. Some sharing formulas are tapered, with the share of "overearnings" that is returned to customers decreasing as the earnings increase. In other cases, just the opposite occurs. Yet others share at a constant rate. Some plans allow the filing of a rate case if the rate of return falls below some level; others do not. Company performance under these alternatives will provide data about the relative effectiveness of the alternative approaches in encouraging efficiencies and innovation.

Variation in the extent of pricing flexibility across states will likewise provide information about the effect of less stringent regulation on market performance. Economists, who advocate that market forces be allowed to operate, will have data to analyze from a variety of approaches to regulating service prices. Perhaps it will be discovered that markets for some services are less competitive than predicted.

As solutions to some problems were formulated, other problems emerged. For example, it was observed that competition among alternative operator service providers for hotel business did not necessarily translate into alternative service providers for the end user. Commissions learned, in other words, that all competition is not created equal. Moreover, what seemed like solutions at one point in time did not work as intended. The New York experiment with rate case moratoria and incentive regulation did not perform without problems. The U S West service-by-service approach to deregulation has been re-evaluated. Other policies are still under evaluation or development. Experience with long-distance deregulation, on the other hand, has generally been favorable.

State regulation is a frequent object of complaint, but it can serve as a convenient ally. While companies groaned about the shackles of regulation, the shackles were being unlocked. Note that the operating companies that made up AT&T did not worry about the incentive effects of rate-of-return regulation— something with which economists have always been concerned—when AT&T was a monopoly provider. State regulation allows for observation of outcome under different policies. It is unlikely that the public would tolerate such differential treatment from a federal agency.

The states have indeed served as laboratories and will continue to do so. Unfortunately, the benefits of these field experiments are difficult to measure, but I believe they are significant. The costs of mistakes at the state level are not as significant as national mistakes would be. As already discussed, a decentralized approach to policymaking can result in greater policy innovation. Whereas some state regulators have chosen to wait for and respond to regulatory reform proposals, several have actively shaped the debate both locally and nationally. We do see competition among jurisdictions. Regulators and industry strive to develop policy that is most appropriate for their states. They seek to learn from the experiences of other states to incorporate features that have desirable effects and avoid including those that do not.

Competitiveness is a concern at both the state and federal levels. Urban areas tend to be served first by state-of-the-art technology. Extension of such service to more rural areas has been slower. Some states have been slow to allow investment in advanced technology in the rate base if, for example, the investment could not be recovered from revenues on an individual switch basis. They have required each switch to recover its costs, averaging other costs over different locations. At the urging of companies and on their own motion, some commissions have recognized that minimum acceptable basic service requires modern technology. Modern switching and transmission are essential for economic growth. Whereas up-to-date technology cannot ensure economic growth, its absence can effectively preclude business development. The availability of current technology is not limited to larger businesses. It is important to residential and small-business customers, as more commercial, educational, and informational activities are accomplished via the telecommunications network. Several state policies have incorporated network improvement and investment programs. State regulators are more likely to be responsive to the particular state's investment needs. For example, regulators in Vermont and Idaho were very interested in the network investment provisions of alternative regulatory plans. It is unlikely that purely federal regulation could incorporate these differences.

It must be acknowledged that each state "experiment" is not independent of others, making accurate policy assessment difficult. Large telecommunications companies may serve as many as 50 or as few as 2 states. Service providers may want their behavior in one jurisdiction to influence regulatory decisions in another. For example, Nebraska has essentially deregulated the provision of most telecommunications services. There, the market actions of U S West may be limited not by the regulatory environment of Nebraska, but by the company's desire to achieve certain regulatory goals in the other states it serves.

The actions of AT&T in one state are likely to depend not only on the regulatory environment in that state, but on the company's regulatory goals in the other states. For example, most state regulators favor the continuation of statewide averaging of long-distance rates. Recognizing this, AT&T is more likely to agree either to regulations that ban deaveraged rates or to voluntarily refrain from implementing such rates.

In summary, the states *are* laboratories. The experiments, however, are not controlled, and the outcomes not always easily quantified, at least over the short term. The lack of independence of company action across the states further complicates assessment of the benefits of state regulation. Yet, state regulation is not without its significant problems: It may impede the competitive process and interfere with the national availability of some services; it is costly; proceedings are often duplicative; and jurisdictional separations of authority are made that are arbitrary from the customer's point of view. For example, most customers think of calling in terms of local versus long distance. It would be difficult for many to understand why intrastate long-distance calls can be more expensive than interstate long-distance calls that travel over longer distances. Variation in policy across states can lead to fragmented investment and planning.

Theory could not answer the question of who should do the regulating. Unfortunately, practice cannot provide a clear answer either. It is easy to list a set of grievances against state and federal regulation and conclude that the system is in need of overhaul. Such a litany would surely elicit defensive responses by the target of such complaints. Responding by pointing out that regulation worked well for 50 years is not particularly helpful, though, when the current industry structure is quite different from what it was 50 years ago. Regulators at both levels are attempting to be responsive. It is not surprising that they have not figured out an optimal strategy, because there is none. As in most policy matters, trade-offs are necessary. Nevertheless, at the state (as well as the federal) level, we see actions designed to encourage efficiency and innovation, facilitate competition, and reduce administrative burdens, while continuing appropriate customer protections.

WHO WILL DO THE REGULATING?

The marbled system of regulation is imperfect, but it is likely to continue until there is greater dissatisfaction with the current system. Frustration will remain with regulatory pricing practices. Government policy involves trade-offs. Efficiency may be the prime concern of economists, but it is only one concern of most policymakers (and likely not of concern at all to others). The industry itself, when the industry and AT&T were virtually one and the same, was largely unconcerned with efficiency. The configuration of prices was an issue that was secondary to promotion of universal service and recovery of the revenue requirement. Regulators have been educated over time as to the importance of cost-based prices. In fact, they have been educated so well that they are now concerned when, under relaxed regulation, service prices exceed costs.

For the telecommunications industry, the transition from regulation to competition continues. It is imperative that jurisdictional battles not impede the development of necessary infrastructure and services. Telecommunications policy is a good example of a public–private partnership. Ownership of the network is private, but many of the services provided are of national concern. Policy coordination is needed. The state hearing rooms should not be used to determine national policy in a fragmented way. For example, the ability of telecommunications companies to provide information services should be determined at the federal level, with input from all concerned, including the states. State regulation and nationally coordinated policy should not be mutually exclusive.

For all intents and purposes, universal service has been achieved. Targeted policies for dealing with the inability of some to afford telephone service have been developed. Localism and diversity, the other cornerstones mentioned by Markey, have also been achieved. Diversity of services and suppliers exist. State regulation has ensured a local approach to and diversity in policymaking. It should be acknowledged that, in recognition of the changed industry structure and the importance of telecommunications to our economy's growth and competitiveness, the laying of new or additional cornerstones may be necessary.

The distinction between local telephone companies and other providers of telecommunications and multimedia services is becoming blurred by corporate acquisitions, mergers, and restructurings. For example, in 1994 Bell Atlantic attempted to acquire Tele-Communications, Inc., the largest cable TV company in the United States: U S West purchased a 25.5% share of Time Warner Entertainment, and NYNEX has been a partner in the takeover of Paramount by Viacom. Ameritech and Rochester Telephone have each put forward proposals that would allow themselves and others to compete as equals in the marketplace. Competitive access providers and others are providing or planning to provide local service.

Until local telephone competition is pervasive, however, telecommunications regulation will be a reality. As competition increases, regulatory reform will continue at both the federal and state levels. It is important to be realistic about just how far that modification can go in the near term. Policy changes as significant as moving to a system of only federal regulation of telecommunications will result from more than the wistful desires of economists or those who are regulated. It must be remembered that acceptability of any change in policy depends on policies already in place. When the airline industry was deregulated, basically one federal agency was involved. We have an established system of state regulation already in place. One prerequisite for significant change is public dissatisfaction.[2] Does that dissatisfaction exist? Is it likely to develop? In the near term, I do not think so, as service quality is perceived to be good and prices for essential services are considered reasonable. The states will continue to be important players in the regulatory arena.

---

[2]Another is a court ruling! I recognize that the public did not clamor for the breakup of AT&T.

A SIMPLE DECISION RULE FOR JURISDICTIONAL ISSUES

John Haring
Strategic Policy Research, Inc.

An important subtext in these analyses of how to draw economically rational jurisdictional boundaries for regulation involves tactical questions of which assignment of authority is likely to lead to efficiency-enhancing regulatory reform. According to Megdal, given uncertainty, differences in consumer tastes, and geographic variation in the extent of and potential for competition, experimentation with a diverse set of approaches in the laboratory of the states is preferred. Egan and Wenders argue that these state laboratories are metaphorically populated by cranks and mad scientists so that, a little paradoxically, the best way to get to truly decentralized (viz., market) decision making is to centralize authority at the federal level and then have federal regulators forebear from exercising that authority.

The latter is a seductive approach, which may, in fact, hold the greatest hope of success. The problem with it is whether federal officials can actually be trusted not to exercise power once they have acquired it. Considerations related to preferences for risk bearing buttress this concern with respect to reliance on a friendly, federal Leviathan (Posner, 1982).

Regulatory capture generally requires an investment of resources to meld a controlling coalition of interests. The costs of forming a coalition vary directly with the size of the relevant polity because costs rise with the number of interests that must be reconciled. This implies that the costs of creating and maintaining an effective coalition will generally be greater at the federal level than at the state level and, therefore, the probability of forming such a coalition at the federal level will be smaller.

On the other hand, intuition suggests that the return from successfully forming an effective coalition will usually be greater at the national level. Expected losses from regulatory capture depend on both the magnitude of the losses incurred in the event of capture and the probability that capture will actually occur. Suppose that expected losses were the same at the state and federal levels of government. Because expected losses at the federal level reflect smaller probabilities of larger losses, the risk averse will attach greater disutility to that outcome than to an actuarially equivalent outcome involving larger probabilities of smaller losses. Risk aversion implies that the relative certainty of smaller losses is preferred to the uncertain (i.e., risky) prospect of potentially large losses. Assuming that citizen-consumers are generally risk averse, decentralization of decision making is preferred.

Megdal identifies a variety of other considerations that argue for state regulation. These include the superior ability of state decision makers to produce and utilize relevant information based on their closer proximity to the people affected by their decisions and the aforementioned desirability for diversity and experimentation where there is uncertainty and given differences in tastes for public goods like regulatory policies (which are nonrivalrous in consumption and nonexclusive). An additional consideration is opportunity for competition among the states and the effectiveness of so-called "voting with your feet" to supply a control on governmental power and behavior. Individual and corporate citizens may be able to shape governmental policies by their decisions to enter or exit a particular jurisdiction. In this way, governments may be constrained by actual or potential competition from other governments for citizens.

The requirements for effective jurisdictional competition bear a close resemblance to those typically posited as requirements for optimal market organization (Bator, 1971; Mueller, 1991). Both require resource mobility, a large number of alternative substitutes, contractual or legislative freedom, and, importantly, the absence of important external effects. Failure of this last condition to be satisfied is the major flaw in the intellectual case for decentralization in markets or governments. Notwithstanding the benefits of competition (jurisdictional or otherwise), there is a clear danger in drawing jurisdictional boundaries (or contracts) too narrowly. In particular, when decision makers take decisions that affect third parties who are not parties to the decision, they may fail to account adequately for effects on these parties.

The economic desirability of internalizing these external or extrajurisdictional effects is the principal rationale for federalization of decision making and presents the fundamental problem for a federal system of governance—how can the preference for decentralized decision making be reconciled with the adverse consequences flowing from policies that ignore important economic interdependencies linking those within and those beyond the boundaries of the decision makers' jurisdiction? Logically, external effects may be internalized by redrawing jurisdictional boundaries so that what was once external becomes internal. The Supremacy Clause of the United States Constitution, when exercised in conjunction with other powers, provides Congress and the Judiciary with a means of redrawing jurisdictional boundaries so that such internalization takes place. Federal regulators exercise this option through their use of preemption.

Internalization of jurisdictional externalities is not a free good. Its costs include the loss of the advantages that more decentralized decision making can bring. These advantages are necessarily forfeited when jurisdictional boundaries are enlarged. But leaving telecommunications policy decisions entirely to state commissions is also unlikely to lead to optimal results. Such action would impose the costs created by each failure to internalize relevant jurisdictional externalities. The important point is that because there are trade-offs between the advantages of both centralization and decentralization, an optimal allocation of decision-making authority may be one in which both the state and the federal governments retain some power to set policy.

In telecommunications, the extrajurisdictional effects that provide the economic rationale for federal preemption of state authority lie principally in economic interdependencies based on complementarities in production and consumption. These complementarities are ubiquitous and occur on both the demand and supply sides of the relevant economic markets.

From an economic standpoint, state actions in regulating intrastate common carriage may adversely affect interstate common carriage, or more generally, interstate commerce. The fact that different equipment can be used to provide interstate and intrastate services or that accounting rules can be fashioned to permit the separation of jointly used equipment in arbitrary ways for regulatory purposes is economically irrelevant. Where there are cost or demand complementarities, actions that affect one side necessarily affect the other. That inputs or outputs are physically, fiscally, or conceptually separable does not alter this fact. Not does it alter the fact that policy rules that fail adequately to recognize the existence of such economic interdependencies will also fail to maximize economic welfare.

Levitz and I (Haring & Levitz, 1989) proposed the use of an "extrajurisdictional effects" test for assignment of jurisdictional authority in telecommunications. Under this test, states would retain all power to regulate intrastate telecommunications as long as their exercise of that power did not impose external effects on persons outside of state boundaries. Under our proposed approach, the FCC would make an initial determination of whether a state regulatory policy related to intrastate common carriage failed to internalize economically relevant (i.e., nonpecuniary) external effects and consequently should be preempted. Appellate courts would review that decision. We believe it would be relatively easy to harmonize this kind of rule with the traditional approach to statutory interpretation and supplied the details of a proposed harmonization in that paper.

Our approach would, in principle, leave much authority to the states and thus potentially leave some scope for experimentation, diversity, and competition among the states, which are the strengths of a decentralized system. At the same time, precisely because telecommunications is an activity where there are liable to be pervasive extrajurisdictional effects, our approach may lead to extensive federal preemption if state power is not exercised responsibly. But that is as it should be if we take externalities and the desirability of internalizing them seriously. When the wires are strummed in one place, they tend to vibrate everywhere else.

INSTITUTIONAL ISSUES

Douglas N. Jones
National Regulatory Research Institute and
Ohio State University

The chapter by Egan and Wenders is an articulate and comprehensive celebration of neoclassical economics as it presumably applies to markets for utility services. It can be added to the sustained drumbeat for the rush to deregulate telecommunications. I believe they carry the argument further than the facts support, they are unfair to state regulation in a number of respects, and their cynicism about public oversight is unwarranted. It is only a slight overstatement to say they argue that public utility regulators should go find gainful employment and get out of the way of the revealed and latent competitive forces in the telecommunications industry, the interplay of which will benignly redound to everyone's benefit. And if some residual regulation must persist, it is better that federal officials do it.

Specifically, Egan and Wenders: (a) decry any court decisions that favor state regulation; (b) dismiss virtually all regulatory decisions as obstructionist and ill-considered; (c) call for relegating commissions to a record-keeping, information dissemination role; (d) ascribe any resistance by state regulators to such a role redefinition as motivated by "loss of personal importance"; (e) worry a great deal about political mischief as harming social welfare with little acknowledgment that corporate mischief that harms social welfare motivated administrative regulation in the first place; and (f) see the impossibility of total deregulation of telecommunications as a political issue with no weight given to the possibility that there might be technical or policy reasons not to do so.

The Megdal chapter offers a more cautious approach and a fairer treatment of what she calls the "marbled structure" of public utility regulation. Measured against the four public policy goals she selects (and assuming away complete deregulation), state commission regulation has a pretty good record. She finds that neither in theory nor in practice can one confidently say that, all things considered, federal regulation is superior to state telecommunications regulation.

Specifically, Megdal argues: (a) state regulators have responded to the need for regulatory reform; (b) in several instances state regulation was clearly ahead of FCC initiatives; (c) the experimental laboratory function of state regulation can be real and not merely a romantic vision; (d) ineffective (experimental) regulation does much less damage at the state than the national level; (e) competitive markets for some telecommunications services may not emerge as rapidly as some predict; and (f) the polity is not calling out for radical change in these institutions in our time-honored dual system of regulation.

I would like to add my own comments based on 35 years of experience in positions related to our system of dual regulation. To argue that there is not a rational division of regulatory authority between the states and counterpart federal agencies is not to say that there is an irrational division. Rather, it is to say (a) there is either a "nonrational" division in the sense that it just evolved without any coherent plan or underpinning of consistent reasoning, or (b) there are any number of divisions that can be argued for with equal or nearly equal sense. There is no single persuasive line of reasoning that leads to a neat "picket-fence" delineation between what should be federal and what should be state domain.

The Constitution does not specify clearly the state role. The Interstate Commerce Act, the Federal Power Act, and the Communications Act contain seemingly clear statements on this matter, but one-sided interpretations, taken together with the federal supremacy doctrine and the invoking of the Interstate Commerce clause, have largely turned these provisions into "double speak." Neither geography (state boundaries) nor function (such as rate design) as bases for dividing up regulatory jurisdiction has stayed the federal reach in the course of blurring the line of demarcation from *Attleboro* and *Colton* to *Narragansett* and *Pike County*. Nor have the questions of which level of government can do it more cheaply, or more efficiently, or more effectively been paramount in deciding jurisdictional issues.

It is difficult not to conclude that the division of authority favoring one arena or the other is mostly ad hoc and mainly dependent on the size of the issue, the inclinations of those sitting on the federal commissions, and general "mood" of the polity regarding centralization, the nature of the technology involved, the congeniality of the courts toward agency expansionism, and the power of the regulated industry to get what it wants. Guiding principles, theory, tight reasoning, or orderly thought have played virtually no role in these changing assignments.

U.S. FEDERALISM

As has been widely noted, the twin developments in recent 20th-century federalism are an expansion of the powers and functions of U.S. governments at all levels and a concomitant redrawing of authority in favor of the federal level.

Consideration of the shift of regulatory jurisdiction from state to federal in recent decades is best done against the backdrop of changes in federal–state relations generally. Specifically, is the alteration of regulatory authority proceeding faster, slower, or at about the same pace as centralizing forces elsewhere in the system? My answer is that telecommunications is proceeding at about the same pace.

The federal–state arguments are familiar. The virtues of federal solutions are uniformity and presumed horizontal equity; the virtues of state solutions are diversity and experimentation. As Justice Brandeis noted in his 1932 dissent to *New State Ice Co. v. Liebman*: "To stay experimentation in things social and economic is a grave responsibility. Denial of the right to experiment may be fraught with serious consequences to the nation. It is one of the happy incidents of the federal system that a single courageous state may, if its citizens choose, serve as a laboratory; and try novel social and economic experiments without risk to the rest of the country."

In the 1980s, much presidential rhetoric seemed to favor a greater emphasis on state authority. But, when the passion of federal regulators for market competition ran up against their administration's predisposition for decentralization, federal deregulation with preemption was chosen every time.

As might be expected, there is a sharp divergence of viewpoints in the current literature on state performance and state policy laboratories. Leach (1970) wrote in *American Federalism*: "Succinctly put, the case against the states is that where action has been required they have too often been inactive; where change has been demanded, they have offered the status quo; where imagination and innovation have been needed, they have seldom come up with satisfactory alternatives" (p. 116). In contrast, Elazar (1974) concluded: "Today there is simply no justification for thinking that the states and localities, either in principle or in practice, are less able to do the job than the federal government" (p. 102).

Applying the themes of state performance, state inventiveness, and state responsiveness to the specific case of telecommunications regulation, I believe most objective observers would award state governmental experiences with high marks.

CAUSES OF CHANGING JURISDICTIONS

One important cause of changes in regulatory authority is the political stance of the regulated firms themselves. Different firms have supported different changes. Most, but not all, telecommunications firms now see FCC policies as more favorable to them.

A second factor is appointments to the regulatory commissions, as who the commissioners are is tremendously important. Appointees who are assertive and expansionist tend to push agency authority to the outer edges and perhaps beyond. When federal commissioners (and staff) with an activist regulatory philosophy (and perhaps a lesser regard for the skills and visions of their state counterparts) feel there is a special mission to perform, such as deregulation, the state–federal jurisdiction boundaries are likely to be altered as opportunities present themselves.

A third factor is the differing role of the Congress in these industries. Congressional action in transportation and energy has never been matched in telecommunications, although there are hints of greater Congressional action in the near future.

Overall, the decline and fall of state regulation has been prematurely implied or predicted a number of times by practitioners and academics.

IMPLICATIONS

For a considerable range, the tension over spheres of jurisdiction can be healthy and constructive. Competition can be a useful force here as well. Eternal vigilance may not only be the price of liberty, but also an imperative if substantial regulatory turf and territory are to remain with the states. Despite all this, the demise of state commission regulation is surely not imminent.

Regulation in a sustained period of rising unit costs, lessened productivity gains, and fewer scale economies is more difficult. So is regulating in a fishbowl with highly skeptical and better informed consumers, many of whom have alternatives that were not available earlier. Deregulation and managing competition within a regulatory framework requires changes in mindsets that must approximate the adjustments we asked of electric and gas utility executives when conservation replaced consumption as the new goal. And, as we are experiencing in telecommunications, the task of managing our way through the transition from a fully regulated national network to a partially regulated, diverse system itself requires a concentration of regulatory authority, even if the result is a net diminution of regulation.

Finally, the combination of large and more expert staffs, together with enhanced budgetary and computer resources, has allowed state commissions to use quantitative analyses and proceed much more rapidly than in the past.

Looking ahead, it is clear that federalism needs to be more than a nostalgic recollection. Given the fact of states, it was not intended that they become merely regional implementation centers and complaint offices for federal government programs. In our excesses toward federal preeminence we seem to have lost our way, and, strangely, we are not engaging in much comprehensive or objective dialogue.

Happily, some recent court cases, such as *Louisiana* have partially restored the state role in telecommunications. Other compromise solutions of continued joint authority are possible and desirable. One possibility is regional, multistate regulation as a halfway house. Another is a requirement that federal agencies issue a "jurisdictional impact statement," akin to an environmental impact statement, that would specify how states would be affected. More broadly, a comprehensive commission on federal–state relations might offer a vision of the future. In any case, it should be possible to craft an ongoing role for state telecommunications regulation that advances and does not interfere with important federal goals in this industry. Just because there may not be "one best way" does not mean that we can not forge a workable interaction, as we have in the past.

ALTERNATIVE PERSPECTIVES ON INTERGOVERNMENTAL RELATIONS

William Gormley
Georgetown University

In telecommunications, as in many other policy domains, responsibility for key policy decisions is shared by the federal government and the states. The relationship between the states and the federal government is complex, messy, fluid, and controversial. There are a number of ways to characterize the current situation. Perhaps the simplest is to say that we live in an era of "regulatory federalism" in which the federal government imposes significant restrictions on state and local governments. These restrictions are evident in the FCC's imposition of a flat rate subscriber line charge despite objections from state governments. They are also evident in a wide variety of policy areas.

THE RISE OF REGULATORY FEDERALISM

The triumph of regulatory federalism in the 1980s is ironic for two reasons. First, the Reagan administration repeatedly stressed the virtue of states' rights, the advantages of state discretion, and the limitations of federal bureaucracies. Despite that rhetoric, the Reagan administration imposed numerous restrictions on state governments when it proved ideologically convenient to do so. In addition, the federal government reduced federal aid substantially, inviting Mario Cuomo to refer to a new era of "fend-for-yourself federalism."

The second irony of regulatory federalism is that the states have become more professional, more capable, and more responsible at precisely the same time that the federal government has placed new restrictions on them (Bowman & Kearney,

1986). State legislatures have become full time, with well-educated professional staffs. Governors' staffs have expanded and chief executives have launched new efforts to achieve administrative integration through reorganization. State bureaucracies have become more professional and the merit system has displaced patronage as the principal mechanism for hiring and promotion. Despite these advantages, federal intrusions continue.

There are, of course, good reasons for federal controls of states. Without federal prodding, the states might stint on welfare payments (Peterson & Rom, 1990). Without federal prodding, the states might stint on environmental protection (Gormley, 1987). In energy, the federal government has a role to play in facilitating energy conservation (a national goal) and in regulating interstate power transactions. In telecommunications, the federal government has a role to play in ensuring uniform technical standards. But where should the federal role stop and the state role begin?

EFFICIENCY AS A CRITERION

In thinking about intergovernmental relations, an economist is likely to begin and end with the criterion of economic efficiency, as defined by Pareto optimality. It is found in numerous treatises on regulatory policy (for example, Kahn, 1970). Some, such as Egan and Wenders, believe that the federal government is more likely than state governments to promote the goal of economic efficiency through deregulation.

I am not convinced that economic efficiency should be the sole litmus test for public policy or for the design of our political institutions. If we lived in a society with a strong welfare state, a guaranteed annual income, and considerable economic equality, one could argue in favor of market solutions, with minimum government interference. If we lived in a society in which there were no subsidies or tax breaks for favored firms, one could make the same argument with a straight face. But we do not.

Thus, in my view it is at least arguable that efficiency should be tempered by other considerations, such as fairness or equity. In telecommunications these issues arise when lifeline rates are being discussed and when the relative burden of residential and business ratepayers is under review. They also arise when the interests of competing firms collide. If economic efficiency were our sole criterion, the way to resolve these controversies would be relatively clear and the case for federal preemption would be equally clear. But in my view other values may also be worth pursuing, such as redistribution, responsiveness, and fairness. The fact that efficiency is more easily operationalized than other values does not mean that efficiency is the superior value!

Another reservation is that federal jurisdiction does not guarantee that economic efficiency will be pursued, even if that were the essential goal. Importunate legislators and meddlesome judges frequently thwart the best bureaucratic efforts to achieve efficiency in regulation. Federal control is not tantamount to policy-making by the FCC. Rather it means decision making by a constellation of actors, including key congressional subcommittee chairs, federal judges, and industry lobbyists. These are not quite the "iron triangles" of legendary fame. But there is no guarantee that their combined "wisdom" will yield an efficient solution.

STATE POLITICS

As we choose between a stronger or weaker federal role in telecommunications, it is important to recognize that regulation is inevitably—and perhaps even appropriately—political. Politics can not, and should not, be banished from the policymaking process. However, there is reason to believe that regulatory politics will differ at the federal and state levels.

At both the federal and state levels there will be pressure to ensure that regulators are accountable for their actions. At the federal level, politics has often taken the form of "upward accountability." Federal regulators have been accountable to politicians and judges. In telecommunications, there is considerable evidence that the FCC has been accountable to the judiciary in recent years, most notably Judge Harold Greene (Stone, 1989). A careful analysis of congressional-bureaucratic interactions would probably reveal considerable accountability to congressional overseers as well (for a model and case study in telecommunications see Ferejohn & Shipan, 1989).

The pattern at the state level, I believe, is rather different. In particular, there is in place an extensive network of "grassroots advocates" and "proxy advocates" who regularly intervene in public utility commission proceedings on behalf of residential consumers. When I studied these intervenors a decade ago, I found that they were very active in three fourths of the states (Gormley, 1983). I also found that they were rather active in telecommunications cases, even though energy cases attracted more media attention at that time.

In my view, the states' greater reliance on grass roots advocates and proxy advocates (downward accountability) has certain advantages over the federal government's greater dependence on judges and politicians (upward accountability). In particular, "catalytic controls" from below preserve a certain discretion for regulators. In contrast, controls from above tend to be more coercive, depriving regulators of precious flexibility (Gormley, 1989).

Of course, there is no guarantee that grassroots advocates and proxy advocates will be either representative, responsive, or responsible. However, regulators can pick and choose, responding to responsible suggestions and deflecting less responsible suggestions. Studies of the impact of public intervenors are at least consistent with this proposition. When I studied grass roots advocates and proxy advocates in the late 1970s I found that they were effective in some areas but not others (Gormley, 1983). Although Teske found no evidence of effective interventions by grassroots advocates in telecommunications cases in the 1980s (possibly because of data limitations), he also found that proxy advocates were effective in some areas but not others (Teske, 1990). Another way of putting this is that public utility commissioners have been selective in responding to interventions by public intervenors, which is probably as it should be.

OTHER CRITERIA FOR CHOICE

In deciding on the proper division of labor between the federal government and the states, it would be mistaken to argue that the same rough division of labor makes sense in all policy domains. There are policy domains where a strong federal role makes sense and policy domains where a strong state role makes sense. Obviously, externalities and economies of scale must be taken into account, as most economists would agree. What are some other appropriate criteria and how might they apply to telecommunications?

One criterion is technical or scientific certainty. What are the consequences of deregulation for competition? And what are the consequences of competition for different types of consumers? Will rates go up? How steeply? If so, will other ratepayers assume a larger portion of the burden? The greater the uncertainty, the weaker the case for federal preemption.

A second criterion is political consensus. Do we agree on the appropriate burden of large businesses, small businesses, and residential consumers? Do we agree on the appropriate role of government in enabling the poor to meet some minimal telecommunications needs, or the "universal service" definition of the future? Do we agree on the appropriate level of subsidies for rural ratepayers? The greater the degree of consensus, the stronger the case for federal preemption.

A third criterion is industry robustness. Are the affected companies large enough to handle a certain amount of regulatory overlap and confusion? Are they adaptable enough to be able to handle significant regulatory change? Are they large enough to lobby effectively in a variety of forums? A strong and vital industry should be able to withstand a certain amount of regulatory diversity.

A fourth criterion is bureaucratic professionalism. Are federal and state bureaucracies both capable of innovation? Are federal and state bureaucracies both capable of identifying efficient solutions, if that is indeed the course they choose to pursue? In answering these questions, it is important to consider not just the size of the regulatory bureaucracy or even how many staff members have advanced degrees. One also needs to know which professions are respresented in what numbers and whether economists in particular have a meaningful role to play.

My view, based on these criteria, is that a considerable degree of state discretion is justified in telecommunications. We have not yet reached a consensus, either on questions of value (politics) or on questions of fact (science). Telecommunications companies are better able to handle overlapping and changing regulations than smaller companies subject to environmental and occupational safety regulations that have caused such controversy. And state bureaucracies, for the most part, are capable of professionalism and innovation, as Cole illustrated in an earlier chapter. Economists, for example, are now well represented on both commissions and commission staffs. Indeed, that has been true for some period of time (Gormley, 1983).

There have been numerous state innovations in telecommunications policy. California adopted a lifeline plan in telecommunications long before the FCC adopted a similar plan. Recent FCC interconnection policies have been guided by the success of state-level policy in New York and elsewhere. The innovative capacity of the states has not yet been exhausted in telecommunications. For all of these reasons, there is much to be said for continuing state discretion.

FEDERALISM AND THE FUTURE

7

THE FEDERAL–STATE FRICTION BUILT INTO THE 1934 ACT AND OPTIONS FOR REFORM

Eli Noam
Columbia University

No other western country besides the United States has a two-tiered level of telecommunications regulation. The sole exception was Canada, until its Supreme Court's *Alberta Government Telephones* decision,[1] which left little in provincial jurisdiction. This makes the United States a minority of one. How can one explain the existence of a two-tiered regulatory structure in U.S. telecommunications?

This is not just an issue of theoretical concern. These two tiers have often been at odds in recent years, and this conflict affects the development of telecommunications policy. So we need to understand the dynamics of state–federal friction.

It is frequently and mistakenly believed that telecommunications regulation by the federal government originated with the 1934 Communications Act. In fact, its antecedents can be traced to the founding of the nation. The U.S. Constitution assigned Congress the general power "To regulate Commerce . . . among the several states" as well as, more specifically, "To establish Post Offices and post roads" (Art. I, Sec. 8). Such authority over interstate commerce and the postal infrastructure provided the background for subsequent federal interest in telecommunications, which was first executed through Congress and the Post Master General.

Federal intervention in telecommunications appeared with the very dawn of telegraphy. In 1843, Congress appropriated $30,000 so that Samuel F.B. Morse

---

[1] *Alberta Government Telephones v. CNCP Telecommunications, CRTC, and the Attorney General of Canada.* (1989) 2 S.C.R. 225.

could build the telegraph line from Washington to Baltimore over which he first demonstrated the practicality of his invention. Within 20 years, the Federal government was regulating telegraph rates under the Pacific Railroad Act of 1862, which provided for Congress and the Post Master General to set rates if railroad and telegraph annual profits exceeded 10% of cost.

In the first years of the 20th century, the states, led by Wisconsin and New York, began exerting their own authority over railroads, energy, and communication. With reasoning that varied from state to state,[2] states began to establish public service commissions and take over telephone regulation from municipal authorities, which had been regulating the new service as part of their control over public rights of way within their jurisdiction.

In 1910, Congress attempted to distinguish federal and state roles in telecommunications with the Mann–Elkins Act,[3] which gave the ICC regulatory authority over interstate telecommunications. The distinction between federal and state jurisdiction was based on the physical test of *interstate* versus *intrastate* segments of communication.

FEDERAL–STATE JURISDICTIONAL AUTHORITY AND THE COURTS

Very shortly thereafter, the Supreme Court decided the so-called *Shreveport* cases,[4] which, although being cases about railroads, had a lasting effect on relations between the states and federal regulators in telecommunications as well. The Texas Railroad Commission had been trying to win competitive advantages for Texas ports by imposing discriminatory railroad rates affecting a rival port, Shreveport, Louisiana. The ICC intervened against the Texas Commission, and the U.S. Supreme Court upheld its action. As a result, the ICC assumed powers over intrastate tariffs where such tariffs had substantial impact on interstate commerce. This became known as the *Shreveport* test, which in the 1920s reduced the role of the states in railroad regulation.

When the 1934 Communications Act was being drafted, ostensibly only to move the ICC's telecommunications jurisdiction to a new, specialized agency without any policy changes, most of the states strongly lobbied for excluding the *Shreveport* railroad standard from telecommunications regulation. And indeed, they won several sections, most particularly Sec. 2(b)(1), which says "nothing in this Act shall be construed to apply or to give the [FCC] jurisdiction . . . for or in connection with intrastate communication service of any carrier." In other words, the pre-*Shreveport* jurisdictional separation was re-established through language that is

----

[2]For example, Wisconsin's PUC emerged in part from Governor Robert La Follette's Progressive reform efforts, and New York established state regulation in part to stave off European-style nationalization of infrastructure sectors. (See Gabel, 1987.)

[3]49 U.S.C 10301 (1982).

[4]See especially, *Houston, E. & W. Tex. Ry. v. United States*, 234 U.S. 342 (1914).

both an exclusionary clause and a rule of construction—a powerful protection to the states that sought it.

The ICC, which was getting out of the telecommunications business, had little incentive to oppose this. The FCC had not been created yet and could not fight for its jurisdictional prerogatives. And AT&T, the giant in the industry, actually liked the provision, as did the smaller independent telephone companies. They understood that the FCC, part of the New Deal "alphabet soup" of activist agencies, began on the left of the political spectrum, that is, focused on redistributional goals and was very much proconsumer. The state commissions had been around for a while, and had traveled further through the life cycle of regulatory commissions, which often includes a period of vigorous youth, followed by maturity characterized by a "subtle relationship in which the mores, attitudes, and thinking of those regulated come to prevail in the approach and thinking of many commissioners" (Bernstein, 1955, p. 78).[5] AT&T, the world's largest corporation, preferred to deal with the state commissions, most of which had no more than a handful of staff to regulate several industries. The comfortable state devils it knew were preferable to the unknown federal ones. Therefore, it was lining up in favor of maintaining or protecting state regulation.

From its perspective, AT&T's concerns were justified. One of the FCC's first actions after it was created was an investigation of the telephone industry, led by Commissioner Paul Atlee Walker. That investigation led to the Walker Report of 1939,[6] which, though not adopted by the Commission, led after World War II to an antitrust lawsuit (1949) that resulted in the 1956 Consent Decree, which in turn begat the next antitrust lawsuit and the AT&T divestiture.

The initial policy divergence between the states and the FCC soon disappeared, however. This happened in part because the FCC was also being gently moved by the interplay of powers into an equilibrium similar to the states'. Thus there emerged from the late 1930s and into the 1970s a remarkable system of *co-regulation*, characterized by a substantial cooperative spirit. The states were mostly in charge of local service; the FCC was mostly in charge of long-distance service. Both were solicitous of AT&T, which steadily extended service throughout the nation at declining real rates and established what was widely recognized as the best telephone system in the world. Moreover, AT&T's financial stability throughout this period made it a model investment for many Americans, creating still another broad constituency in favor of the status quo.

This structure started to program its own decline when it began to draw on long-distance service to subsidize local residential rates. The decision was made by state regulators, shortly after World War II, partly in response to political

---

[5]For other perspectives on capture or life-cycle theories, see for example, Stigler (1971, 1975), Posner, (1974), Peltzman (1976), Edelman (1964), and Olson (1982). For earlier critiques that also analyzed the ties between regulators and the regulated, see Huntington (1952) and Jaffe (1954).

[6]Federal Communications Commission, *Investigation of the Telephone Industry in the United States*, H. Doc. No. 340, 76th Congress, 1st Sess., Washington: GPO, 1939.

pressure. U.S. Senate Majority Leader Ernest W. McFarland (D–Arizona), who also chaired the Senate Communications Subcommittee, was the spearhead. The trend was supported by AT&T, which saw it as part of an implicit bargain in which it would hold residential rates low through cross-subsidies from long-distance service in return for entry barriers to potential competitors.[7]

At the same time, the policy of "universal service" was pursued throughout the country, in which networks connected new subscribers beyond the point of purely economic equilibrium. Redistributory mechanisms took hold through which the majority of the network users, via political means, extracted a subsidy from a minority of the network users. As redistribution grew, some users wanted to leave the network, at least for part of their communications needs, if legally permitted to do so. As this process developed, it led to the rift between the states and federal officials.

Federal regulators were more amenable to local competitive entry because the long-distance rates for which they were responsible were declining. The federal regulators had the rare privilege to preside over an industry segment whose prices dropped as performance rose. They were also at a comfortable distance from any grassroots discontent, in contrast to state regulators, who in some states are elected and generally are held more directly responsible for telephone rates. As a result, the FCC, which started out during the New Deal to the left of the states, moved to the right of the states.

The rift first opened over the interconnection of terminal equipment not supplied by the Bell system. Many states, led by North Carolina, opposed the connection of subscriber-owned terminal equipment until they were decisively rebuffed by the courts,[8] which held that the states' jurisdiction was limited to local services and facilities, as well as matters "that in their nature and effect are separable from and do not substantially affect the conduct or development of interstate communications."[9] Because the court's decision concluded that even the handset in the customer's home affected interstate communications, it in effect mooted the 1934 Act's separation of intrastate and interstate that had been the legal linchpin of the cooperative system.

JURISDICTIONAL IMPACT OF CHANGES IN NETWORKS

When the AT&T network monopoly began to disintegrate and rivals emerged, their interconnection with the traditional network became essential, which has

---

[7]See, for example, deButts' (1973) speech to the National Association of Regulatory Utility Commissioners, which argued that competition in the telecommunications sector "cannot help but in the long run hurt most people," by destroying the system that allowed monopoly providers to furnish dependable, economical service (quoted in Coll, 1986, p. 40).

[8]*North Carolina Utilities Commission v. FCC*, 537 F.2d 787 (4th Cir. 1976); *cert. denied*, 429 U.S. 1076 (1976); 552 F.2d 1036 (4th Cir. 1977), *cert. denied*, 434 U.S. 874 (1977).

[9]537 F.2d at 793.

two implications. First, through the addition of interface points, the network over time becomes increasingly modularized. Second, new entrants begin cream-skimming, that is, they attack the above-cost segment. Interconnection arrangements were established, for example, in the *Carterfone*[10] decision that allowed subscriber-owned terminal equipment to connect to the network, and the *Execunet*[11] decision, which allowed long-distance carriers to interconnect into the local loop of the traditional network.

Modularization and interconnection are not just of historical interest. Modularization will inexorably continue and will have profound implications on federal–state relations. At the simplest level, the states have opposed many interconnection arrangements such as those just described because they identified their interests with those of the monopoly.[12] The myth is that in the *North Carolina* case, for example, the states were not really opposed to interconnection of terminal equipment but were really fighting for states' rights. This is reminiscent of the argument that the Civil War was fought primarily over a procedural—the scope of states' rights—rather than over a substantial policy disagreement. In telecommunications, there was also a substantive policy disagreement. States believed that local telephone rates could be kept down by the contribution from equipment profits. This was AT&T's argument. From the consumer's perspective it was flawed, but many state regulators accepted it.

As the modularization of the network increases, ever greater parts of telecommunications service will be composed of multiple blocks or modules. As a direct consequence, notions of interstate and intrastate services will blur because the component modules of each service will cross jurisdiction: some of them will be interstate, some of them will be intrastate, some of them will be international, and others will exist nowhere physically. Networks are becoming relational, not locational.

The traditional notion of jurisdictional separation found in the 1934 Act was based on a linear, spatial concept of what a network was, borrowed from earlier railroad regulation: local was close, long distance was far, international still farther. This was based on network architecture, which was in turn based on technology and economics. Networks were largely configured to minimize transmission distance. But today, transmission has become a much smaller portion of telecommunications costs and will continue to decline, making telecommunications relatively distance insensitive. As a result, the nature of the network architecture changes, with a series of consequences for the jurisdictional question.

Network modularity and interconnectivity affect not only transmission, but also switching, including local switching, which traditionally was the essence of

---

[10]*Use of Carterfone Device*, 13 FCC 2d 420 (1968).

[11]*MCI Telecommunications Corp. v. FCC*, 561 F.2d 365 (D.C. Cir. 1977), (*Execunet I*); see also *MCI Telecommunications Corp. v. FCC*, 580 F.2d 590 (D.C. Cir.), *cert. denied*, 439 U.S. 980 (1978), (*Execunet II*).

[12]See the discussion of *North Carolina Utilities Commission v. FCC*, op. cit., *supra*.

intrastate jurisdiction. The FCC's *Arco* decision, which allowed users to interconnect to the local exchange of their choice as long as it is "privately beneficial without being publicly detrimental,"[13] marked a significant step toward breaking the grip of state jurisdiction on switching, even though it received little attention. The FCC's decision, which in effect permitted one telephone company to interconnect into another telephone company's central office, suggests that just as one can plug a "Mickey Mouse" telephone or a PBX into the network, one can also plug an entire network into the network. And although in this instance it was one Texas-based LEC versus another (Southwestern Bell vs. GTE), there is no reason why interconnection on this scale could not occur across state lines. Once that happens, local switching may just as easily be interstate as intrastate.

Another issue arises with the emergence of private networks, first for large users and user groups such as banks, universities, manufacturers, and their suppliers and dealers. This means that a network ceases to be a territorial concept and becomes a group concept. It becomes a functional rather than a spatial arrangement. The concept of "intrastate" will become a relatively meaningless concept in this environment. For example, the interconnection of stock markets was begun as merely an arrangement to facilitate data transfer. But in time, with computers talking to computers located everywhere in the world, the physical location became meaningless. The network has become the market, and the market exists in no physical location. As the notion of the "New York market" loses its meaning, so does that of the "New York network."

This evolution can also be moved beyond group networks. Just as we now have personal computers, which only two decades ago was a concept that people did not anticipate, we can also think of personal networks in the future. These are custom-tailored, individualized networks that are configured along the lines of individual needs. As we look ahead, then, we can expect arrangements that do not fit the traditional notion of what a network looks like and the jurisdictional basis for its regulation.

The increasing importance of software also contributes to the diminished locational aspect of telecommunications networks. It is very difficult to say exactly where network software is physically located. For example, software can mean control functions that interact and that are distributed. It can mean interaction of databases, programs, and processes undertaken at locations that are far apart from each other.

Another prospect is that the fiber-based fast-packet-type network of the future may create a kind of "fiber ether." Communications will no longer move on a point-to-point line but along an increasingly dense matrix through which the information routes itself. This would be the case in a fast-packet environment such as envisioned in SONET. In such an arrangement, it would become impos-

---

[13]*In re Atlantic Richfield Co., Memorandum Opinion and Order*, 3 FCC Rcd 3089 (1988) at 3091, *aff'd sub nom., Public Utility Comm'n of Texas v. FCC*, 886 F.2d 1325 (D.C. Cir. 1989).

sible to determine how the information moves, much less whether it moves interstate or intrastate. In fact, it can be both at the same time, with part of the information in a single call moving one way and part of it moving another way.

What are the implications of these technological changes? One is that in the future the core of identifiably intrastate communications activities will shrink continuously. Furthermore, the share of communications activities generally that are regulated by anybody is going to shrink. The net result is a shrinking share of a shrinking share, which means that state regulation will be under continuous pressure.

In this evolution, the courts, until recently, have given the FCC most of what it wanted to deal with changed circumstances. For example, the North Carolina case referred to earlier established an "adverse effect test." The court said, in essence, that if the FCC believed a state's action was adversely affecting federal regulation, the court would support the FCC.

However, the courts have in recent years become generally more involved in resolving telecommunications policy disputes and significantly more likely to overturn FCC decisions in recent years. In the first 35 years after the Communications Act was passed, 37 common carrier decisions were reviewed by the circuit courts. Of those 37, only 5 (16%) were reversed or remanded. From 1970 through 1978, a period during which the FCC began opening the markets for network equipment and long-distance service, these same courts reviewed 61 common carrier decisions and reversed or remanded 14 of them (about 23%). As deregulation took hold and the network disintegration process outlined earlier gained momentum from 1979 through 1989, the courts took on 129 of the Commission's common carrier decisions (an average of one every month) and decided against the FCC in 38 cases, or 30% (Blau, 1990).

Against that background, it is not surprising that in several cases (e.g., *Louisiana*,[14] *California*,[15] *NARUC III*[16]) courts have pulled back from the adverse effect test that presumed in favor of the FCC. This change is a result of a coalition of results-oriented liberals and states'-rights oriented traditionalists. Conservatives had always had a dilemma on the issue of preemption, as they favored both deregulation and states' rights. Both are conservative values, but they tend to conflict when the FCC tries to impose deregulatory rules on the states. The Reagan administration, having to make the choice, picked deregulation as the priority and left states' rights as less important when the chips were down. Some conservative judges did, too. For example, Warren Burger, when he was still a circuit court judge, upheld federal jurisdiction, writing: "Any other determination would tend to fragment the regulation of a communications activity which cannot be regulated on any realistic basis except by the central authority. Fifty states

---

[14]*Louisiana Public Service Commission v. FCC*, 476 U.S. 355 (1986).
[15]*People of the State of California v. FCC*, 905 F.2d 1217 (9th Cir. 1990).
[16]*NARUC v. FCC*, 880 F.2d 422 (D.C. Cir. 1989).

and myriad local authorities cannot effectively deal with bits and pieces of what is really a unified system of communication."[17]

However, his successor, Chief Justice William Rehnquist, has been considerably more states'-rights oriented, and Justice Brennan, one of the Court's longtime liberals, also took a very "original intent" approach to the Communication Act's jurisdictional separation in the *Louisiana* decision.

What is thus emerging, in particular in the *California* case, is a return to a very literal reading of the 1934 Act. And this as if the underlying environment has not changed dramatically: In 1934, 98% of all calls were intrastate and all the new telecommunications services and network architectures previously described did not exist. The environment has changed radically, yet the courts are returning to a very literal reading of what is intrastate and what is interstate.

POSSIBLE SOLUTIONS

This raises the question of what the right way to deal with this problem may be. There are several types of nonconflicting regulation.

Total Deregulation.   Eliminate regulation and the jurisdictional problems disappear, but it is unlikely that total deregulation is feasible for some time. In a partly competitive system with bottlenecks, numerous functions remain for expert agencies to ensure the functioning of a pluralistic network. For example, interconnection issues such as financial charges and content access may require a regulatory arbitrator.

Exclusive Federal Regulation.   The concept of a regulatory "czar" is based on an essentially romantic notion that one decisive person or agency might be able to overcome all of the contradictions in the various pulls and pushes of society.

A variant is:

Exclusive Federal Jurisdiction With Some Regional Variation.   This might entail an arrangement like the field offices of the FCC that would differentiate, for example, between the more rural U S West region and the more urban NYNEX territory.

Exclude Federal Jurisdiction Altogether.   This is the other extreme, and it creates problems of an absence of cohesion, particularly now that the Bell system, which used to provide uniformity standards nationwide, has been splintered.

---

[17]*General Telephone Co. v. FCC*, 413 F.2d 390 (D.C. Cir.), *cert. denied*, 396 U.S. 888 (1969).

If one looks at the jurisdictional issue in a detached way, as illustrated in chapters 5 and 6, one can find the benefits derived from both federal and state involvement. For example, to have only federal jurisdiction would mean that the federal government would have total regulatory rights over local franchising. Federal regulators would control the way in which local governments would grant rights of way. Exclusive federal jurisdiction would also forfeit the "laboratory" function that the states now serve and states' ability to tailor policies to local conditions and traditions. Yet to have only state regulation would lead to needless duplication and contradiction in regulatory efforts, and would forfeit the benefits of national integration.

There is thus a logic to co-regulation, though it must be much improved. Here, the options include:

The Status Quo.   That may well be the way things will turn out because the existing system is very hard to change. Too many interest groups have a stake in the status quo, and advocates of change must outnumber traditionalists by a wide margin for policy to actually change. But the lesson from recent history in Eastern Europe is that institutions that are not responsive or capable of changing themselves sooner or later will be in trouble. If the system does not work well, it will eventually be changed. This applies to telecommunications regulation as well.

Clarify or Modify the Respective Regulatory Spheres.   This is an attractive notion: Perhaps we can draw the line more brightly, and possibly shift it a bit in light of changed circumstances. However, it is unlikely that this can be done in a way that would be superior to the present interstate–intrastate separation. If one chooses to assign state or federal jurisdiction based on, for example, functional distinctions, new problems would emerge immediately. For example, if consumer complaints become solely a state regulatory responsibility, then questions immediately arise about the applicability of federal consumer protection law. Rather than separate the spheres, the only result is to shift the focus of the dispute to other lines.

Joint Regulation.   This option would entail a body, such as a joint board, consisting of federal and state representatives, which would have full authority. That may be a good idea in theory, but in practice, everything would depend on the composition of the board. If there are more state representatives than federal ones, state prerogatives would likely prevail. Under the reverse scenario, the federal side predominates. Therefore this approach does not seem to solve anything.

Broad Federal Rules and Authority That the States May Implement Flexibly Within Well-Defined Parameters.   This arrangement may require the modification of the preemption rules. For example, a test could be established that gives the FCC authority to define issues of national jurisdiction, but assign it a burden of proof that nonconforming state regulation would create substantial harm to specified national policies and goals. Such a test would enable the FCC to create a unified national policy, still allowing state experimentation and responsiveness to local needs as long as the state does not actually contravene federal policy. In some instances, this may mean an expansion of federal jurisdiction into state territory, but the reverse is also true. States could regulate some interstate services as long as they do not negatively affect stated national policies and goals. This would substitute a test of nationally necessary policy standards versus decentralized rules for the increasingly unworkable interstate–intrastate separation, and reflects the necessity for national policy as well as the possibility of having local policy. As a result, the focus would stay on achieving policy goals rather than maintaining legal distinctions.

I have argued that the technical-economic trends are against the states because the core of what is intrastate will continue to shrink, regardless of what the courts may say. Thus, even if one fears that this option would in some instances reduce the authority of the states, this is going to happen anyway.

This proposal would in effect establish several regulatory ranges. First, the one assigned by law and preemption to the FCC; second, one delegated by the FCC for state treatment, subject to its broad overall rules; third, those areas not meeting the preemption criteria for the FCC, which would be state regulated; a fourth area may be established for new issues for which there is no national policy determination.

How does one get there from here? The first alternative is to amend the 1934 Act. For example, a subclause could be added that, after describing the role of the states and of the FCC, would specify its ability to preempt, its ability to delegate, and the states' ability to experiment. But the problem with any legislation is the ability of interested parties to block change. Even more important questions do not get resolved, so there is little room for optimism that an amendment to the 1934 Act would be addressed. If anything, Congress would likely give itself more powers as the result of any rewrite, as may occur in 1994.

Focus on Developing Better Federal–State Relations. The final option is essentially cooperative. In the past, the FCC was not adept in keeping relations with the states positive. On the other hand, state regulators created a common denominator based on a "solidarity of the oppressed." Too often this has resulted in a knee-jerk opposition.

As for the FCC, it must affirm the value of state experimentation. Such experimentation would get states to be more evenly distributed around the FCC's pole along the axis described earlier, which is closer to the ideal in a federal system. Under those conditions, national policy stands roughly at the center of state policies, which indicates that it has widespread acceptability. Such conditions yield policy diversity although not fundamentally different policies, because by definition state policies will be, on the average, similar to the federal one. As a result, the choice of regulatory forum would be less outcome determinative than it is now, when most states fall to the left, or redistributory side, of the FCC. Such an outcome would also benefit the states, who would find that because their policies were more evenly distributed around federal policy, there would be far less incentive for preemption by federal regulators. The process has been quietly taking place for the past several years.

A cooperative federal–state approach will also be crucial to address equity or redistribution issues in the future, because it will be hard for states to do it alone. Under the old monopoly system, redistribution to keep local rates low was achieved internally through cross-subsidies from long-distance service, described already as part of the evolution of networks under pressure for universal service. Under increasingly competitive conditions now emerging in the telecommunications sector, it will become difficult to generate money for such subsidies. Competition normally has the effect of driving rates down toward marginal costs, thereby eliminating the source of cross-subsidies. Indeed, as experience has demonstrated, the first areas in which competition emerges is for the above-cost segments of the industry, especially long-distance service. These conditions dry up the traditional sources for redistribution. In light of these problems and realities, the only way to move forward is to substitute an explicit form for the traditional internally generated and ad hoc contributions. Yet for any state to initiate this would lead to an out-migration of communications business and traffic. Thus, any explicit charges need to be implemented, if at all, on the federal level, with the revenues then distributed to the states for use as they see fit, according to their priorities.

In other words, it is necessary to establish again a cooperative model and realize that in the era of the network of networks, the maximization of jurisdictional spheres is a game for bureaucrats, not for policymakers. There is probably no longer a single optimal locus or size for a jurisdiction. Some issues are local, others national, others state, and others regional—it depends on the issue. There is no reason to believe that this is static. Napoleon created a system of administrative "departments" based on the distance that a man on horseback could cover in one day. The emergence of powerful telecommunications has probably made the optimal jurisdictional size much larger. In that sense, states should not mourn some loss, over time, of jurisdiction over some telecommunications functions, but rather understand that this reflects their success in creating ubiquitous and powerful communications media. There is plenty of work left for everybody.

LEGAL ISSUES IN PREEMPTION

Henry Geller

Markle Foundation

The main thrust of my comments is simply stated: State regulation of intrastate telecommunications is desirable on a number of grounds—the so-called "grass roots" factor (i.e., the states are closer and more attuned to the particular facts in their jurisdictions than a centralized federal authority can be), and Justice Brandeis' apt point about the states as "laboratories" (i.e., the gains from substantial diversity in policy approaches among the states; Haring & Levitz, 1989). Thus, some states have been more innovative than the FCC in deregulatory approaches, such as with respect to intrastate toll and in substituting price regulation for the traditional rate-of-return method.

But there is a clear need for a federal "captain" in several areas, such as deregulation of enhanced services or effective introduction of new radio-based services. Congress should supply general guidelines, but has failed to do so, except for the 1993 legislation that preempts state rate regulation of cellular and other radio-based services. That leaves the task to its delegatee, the FCC, but as the 1990 Ninth Circuit decision shows,[1] the FCC is greatly handicapped here because of the existence of an anachronistic provision of the 1934 Act. As it confronts other telecommunications issues in 1994 and beyond, the Congress should act to repeal the provision and allow the normal preemption test to prevail in this area. Congress is unlikely to do this. Thus, as Noam noted, we are likely to muddle along, with strong cooperation between the states and the federal

---

[1] *People of California v. FCC*, 905 F.2d 1217 (9th Circuit 1990).

regulators as the best hope for limiting potential damage in this area, so important to national growth.

PREEMPTION STANDARDS

First, there is the matter of the normal preemption approach in our federal–state system. This can be gleaned by looking at court decisions in the cable television area. Thus, in *City of New York v. FCC* (486 U.S. 57, 63-64, 1988), sustaining FCC preemption of state technical standards governing cable television, the Court first pointed to the Constitution ("the Laws of the United States . . . shall be the supreme Law of the Land"), and then notes that this phrase "encompasses both federal statutes themselves and federal regulations that are properly adopted in accordance with statutory authorization." And in *Capital cities Cable, Inc. v. Crisp* (467 U.S. 691, 699, 1984), where again federal preemption of state cable regulatory action was found to be valid, the Court stated that enforcement of a state regulation may be preempted by federal law, *inter alia*, "when compliance with both state and federal law is impossible or when the state law 'stands as an obstacle to the accomplishment and execution of the full purposes and objectives of Congress' . . . ." The court further quoted:

> Federal regulations have no less pre-emptive effect than federal statutes. Where Congress has directed an administrator to exercise his discretion, his judgments are subject to judicial review only to determine whether he has exceeded his statutory authority or acted arbitrarily. When the administrator promulgates regulations intended to pre-empt state law, the court's inquiry is similarly limited. "If his choice represents a reasonable accommodation of conflicting policies that were committed to the agency's care by statute, we should not disturb it unless it appears from the statute or its legislative history that the accommodation is not one that Congress would have sanctioned."

These standards—reasonable accommodation of conflicting policies and especially, "obstacle to the accomplishment of the full [federal] purposes and objectives"—clearly bestow wide authority on federal regulators to act preemptively, unless these is a specific Congressional bar. As a final example, there is the Brookhaven case,[2] where the court tersely held that the FCC "has the authority to preempt state and local price regulation of special pay cable programming" because its "policy of permitting development free of price restraints at every level is reasonably ancillary to the objective of increasing program diversity."

In the telephone area, no problems arose for decades simply because federal and state regulators agreed on policy aims and the means for achieving those

---

[2]*Brookhaven Cable TV, Inc., v. Kelly*, 573 F.2d 765, 767 (2d Circuit 1978), *cert. denied*, 441 U.S. 907 (1979).

aims; in particular achieving universal service by fostering low rates for local service (with the states' benefitting greatly from shifts of costs to the interstate jurisdiction where costs were declining because of technological advances). But this extended honeymoon ended in the 1970s when federal actions introduced competition into the system and thus threatened to undermine state maintenance of low rates through the subsidy scheme. This led to a series of court cases testing whether federal policies, based on Sections 1 and 2(a) of the Communications Act, should prevail over state actions taken under Sections 2(b) and 221(b) of the Act.

The FCC initially won several large battles. Thus, in two Fourth Circuit cases,[3] the court upheld an FCC order preempting state regulations prohibiting subscribers from connecting *their own* phone sets to any telephone facilities used for intrastate calling, because limiting sets to either interstate or intrastate use was a "practical and economic impossibility," thereby rendering "federal tariffs authorizing interconnection . . . nugatory" (*NCUC II*, 552 F.2d at 1043). The high-water mark here is the FCC action in Computer II,[4] where the FCC, acting under its ancillary jurisdiction in Sections 1, 2(a), and 3(a) of the Act, deregulated all but the basic service market in order to enhance consumer choice and encourage efficiencies, and, to achieve fully these objectives, preempted the states from regulating the offering of CPE and enhanced services. In a sweeping decision, the D.C. Circuit affirmed these preemptive actions.

But there has always been a bomb waiting to explode in the Communication Act's provisions governing this issue. In 1934, when power over interstate communications services was centralized in the FCC, Title II incorporated provisions of the Interstate Commerce Act. This, in turn, raised concerns of state regulators that their intrastate actions could be readily preempted by the new agency based on the 1914 *Shreveport* decision,[5] holding that the ICC could preempt an intrastate railroad rate prescribed by the state in order to prevent unjust discrimination against interstate traffic. The states therefore sought and obtained a provision in Section 2(b) (and 221(b)), which preserved their authority over "charges, clarifications, practices, or regulations for or in connection with intrastate common carriage." Although the legislative history makes clear that this provision was added because of concern over *Shreveport*, the Act itself sets up a puzzling conflict: Sections 1 and 2(a) give the FCC exclusive authority to regulate interstate communications; Section 2(b) vests similar exclusive authority to regulate intrastate communications in the states; and nowhere does the Act come to grips with the stark reality that the same telephone plant is used to carry both interstate and intrastate traffic.

---

[3]*NCUC I*, 537 F.2d 787 (4th Circuit 1976), *cert. denied*, 429 U.S. 1027; *NCUC II*, 552 F.2d 1036 (4th Circuit), *cert. denied*, 434 U.S. 834 (1977).

[4]See 77 FCC 2d 384 (1980); *recon.*, 84 FCC 2d 512; *further recon.*, 88 FCC 2d 512 (1981), *aff'd sub nom. Computer & Communications Industry Ass'n v. FCC*, 693 F.2d 198 (D.C. Circuit 1982), *cert. denied*, 461 U.S. 938 (1983).

[5]234 U.S. 342.

With the growing conflict between the federal government and the states, a case was bound to reach the Supreme Court eventually, and in the 1986 Supreme Court decision in *Louisiana*,[6] the bomb went off. The Court did not resolve all conflicts. Rather, it held that Congress, in dealing with the *Shreveport* issue, meant what it said in Section 2(b) and that there are accordingly "two hands on the wheel." The Court disapproved the FCC's attempt to preempt states from applying their own depreciation rules in setting intrastate rates, even though the FCC had found that such state rules frustrated the federal policy served by its own depreciation rules regarding this same equipment. The Court found that Section 1 does not permit FCC preemption of any state regulation encroaching on interstate communication because of the express provision in Section 2(b); that provision, it held, does not bar FCC action only when the matter in question is purely local and does not affect interstate communication. Rather, Section 2(b) explicitly limits the FCC's power by "fencing off from FCC reach or regulation intrastate matters—indeed including matters 'in connection with intrastate service' " (476 U.S. at 370). Because "it is certainly possible" for interstate and intrastate depreciation rules to coexist, even though it might adversely affect full accomplishment of the federal purpose, the Court set aside the FCC preemption order. The Court, citing the *NCUC* cases, did recognize that the FCC may preempt conflicting states rules where it cannot "separate the interstate and the intrastate components of [its] asserted . . . regulation" and thus "state regulation would negate the federal tariff" (476 U.S. at 374).

Louisiana thus established an entirely different standard from that normally applied in preemption cases. Even if the FCC has reasonably concluded that the state regulation prevents full accomplishment of the federal objective, it cannot preempt. Rather, it must find that there is no way to separate the intrastate and interstate components and that the state regulation *negates* or renders nugatory the federal action. This infeasibility/negation test is obviously a much more difficult hurdle for the federal agency.

Louisiana has profoundly affected subsequent FCC and court decisions. Thus, the FCC retreated from a prior preemptive ruling as to state regulation of cable companies' institutional networks because of its adverse effect on interstate competitive markets.[7] The D.C. Circuit held that the FCC can not preempt state regulation of "inside wiring" used for both interstate and intrastate purposes, on the ground that installation and maintenance of such wiring is not a common carrier service, because Section 2(b)(1) includes matters "in connection with" intrastate service.[8] The Court further held that "a valid FCC preemption order

---

[6]*Louisiana Public Service Commission v. FCC*, 476 U.S. 355 (1986).

[7]See *Cox Cable Communications, Inc.*, 1 FCC Record 561 (1986), *vacating as moot*, 102 FCC 2d 110 (1985) ("state regulation of institutional services offered by cable companies that acts as a *de facto* or *de jure* barrier to entry into the interstate communications market or to the provision of interstate communications must be preempted").

[8]See *NARUC v. FCC*, 880 F.2d 422, 426 (D.C. Circuit 1989).

must be limited to [state] tariffing that would necessarily thwart or impede the operation of a free market in the installation and maintenance of inside wiring" (*id.* at 430). On remand, however, the Commission, "in the interests of comity, . . . determined that it should monitor any state actions in relation to the prices and terms and conditions of service under which telephone companies provide [unbundled] inside wiring services, rather than propose to preempt such action at this time."[9] This reliance on a "lifted eyebrow" technique instead of preemption undoubtedly stems from caution due to *Louisiana*.

The same caution is apparent in the FCC's 1988 *Open Network Architecture Order*,[10] where the Commission declined to preempt state actions conflicting with the tariffing and technical requirements of Computer III (*id.* at 168). Rather to reduce conflict between federal and state ONA policies, the Commission created a state–federal ONA conference (*id.* at 115).

The lower appellate courts also have adhered to the teaching of *Louisiana*. Thus, in the *ARCO* decision,[11] where the FCC, in order to protect a customer's "federal right of interconnection" preempted conflicting state regulations embodying the state's determination that local service be provided by the state-franchised monopoly, the preemption was sustained because on the record before the agency, it was concededly not possible to "separate the interstate and intrastate components" of the state regulation and therefore deferring to the state regulation would negate the federal policy on interconnection for interstate purposes.[12]

In two other decisions, the courts displayed the wide scope of *Louisiana*. In *California v. FCC*,[13] the FCC, acting under its Title III authority over radio licensing, preempted state regulation of wholly intrastate radio common carrier services provided on FM subcarrier frequencies, to the extent that such state regulation blocks or impedes entry of these services.[14] For if that were the case, the FCC pointed out that the state action would conflict with the public interest licensing determination of the FCC, restrict the beneficial use of the radio spectrum, and frustrate the FCC's efforts to encourage competition. The court noted that the FCC had made "a persuasive case in support of its policy objectives, but that case must be made to Congress and not to this court," in light of the holding in *Louisiana*.[15]

---

[9]FCC Report No. DC-1645, May 31, 1990.

[10]4 FCC Record 1 (1988).

[11]*Public Utility Commission of Texas v. FCC*, 886 F.2d 1325 (D.C. Circuit 1989).

[12]*Id.* at 1333-5. The court therefore found "it is unnecessary in this case to accept the broad proposition that a private microwave operator has an absolute federal right of access to the public switched network at location of its choice, unimpaired by state regulatory interests, in order to affirm the FCC's order." *Id.* at 1335.

[13]798 F.2d 1515 (D.C. Circuit 1986).

[14]57 R.R. 2d 1607. *recon. denied*, 57 R.R. 2d 1684 (1984).

[15]In one of Congress' few definitive actions in this area, it did preempt state rate regulation of cellular and other wireless services in legislation in 1993 related to auctioning off portions of the electronic spectrum.

Finally, the most devastating setback to the FCC was the Ninth Circuit's decision in *California v. FCC*,[16] setting aside the FCC's *Third Computer Inquiry* decision. The FCC there held that it would preempt state regulations conflicting with its nonstructural safeguard. The court reversed, holding that under *Louisiana*, states retained the right to require separate subsidiaries for wholly intrastate enhanced service operations by carriers. Even more important, because the FCC foolishly reopened the issue of federal preemption of state enhanced service regulation, the court held that under *Louisiana* states could regulate the terms, charges, and conditions of such operations. And two jurisdictions, Florida and the District of Columbia, soon indicated they would explore such regulation, and the FCC opened proceedings to deal with the remand order.[17]

IMPLICATIONS

It is the poorest possible policy for states to regulate the charges or terms of enhanced services. Consumer Premises Equipment (CPE) is completely deregulated, and CPE and enhanced services are functionally equivalent. Further, enhanced services offered by noncarriers are not regulated, whereas those offered by carriers can now be subject to regulation—again an anomalous situation. The plain fact is that regulation in an area where there is effective competition makes no sense. Further, if the states do impose regulatory conditions on important enhanced service providers such as the LECs, it can impede the nationwide development of such services, so important to the United States in this information age and era of global competition.

But the crucial consideration for my purposes is not the substantive one, but rather the extraordinary absence of any federal captain in this most vital area of telecommunications. The normal preemptive standards apply and there is thus a federal captain in the broadcast area, in cable television, in Master Antenna services (MATV),[18] and in Satellite Master Antenna Television (SMATV).[19] But in the most critical area of all—telecommunications—the issue is ruled by a 1934 provision, based on restricting the application of a 1914 Supreme Court decision. There is no clear-cut federal authority to ensure full effectuation of important national goals. For example, important ONA goals can be thwarted by state action. The FCC's deregulation of CPE, accomplished by preemption of conflicting state action, has been a resounding success for the nation. But if the

---

[16]905 F.2d 1217 (9th Circuit 1990).

[17]See, for example, *Telecommunications Reports*, November 5, 1990, at 2; December 17, 1990, at 1–6.

[18]See *New York State Com. v. FCC*, 669 F.2d 58 (affirming FCC's action preempting state's application of cable franchising conditions to MATC systems).

[19]See *New York State Com. v. FCC*, 749 F.2d 804 (D.C. Circuit 1984) (affirming FCC action preempting state's application of cable certification requirements to SMATV).

states so wish, they can undermine the similarly desirable effort in the functional equivalent to CPE, the enhanced services.

This is folly. No other nation operates in this fashion. Again, I stress that the United States, in light of its size, gains immensely from state regulation (the grassroots and laboratory factors discussed initially). But there is a need for federal leadership and preemption when the national interest so requires—when state action has perverse effects beyond the state's borders. There is simply no warrant for handicapping the federal captain (the FCC) in this area.

The problem would not be so serious if Congress itself acted to provide the necessary national structure or guidelines, on appropriate occasion. The preemption of state cellular rate regulation in 1993 was a start. Although Congress will hopefully continue this line of activity, based on the experience of the last two decades, one must be uncertain. Congress does hold hearings and send messages, but because of the strong conflicts among contending parties and industries, the Act remains in the same form, echoing 1934 and 1910 provisions, despite the enormous changes that have occurred since 1970. In my view, Congress should at least revise the Act so the normal preemptive standard applies to this area.

In her final speech as NARUC president at the Annual Meeting on November 12, 1990, Sharon Nelson deplored the absence of leadership by the federal government in the last decade, and called for that government "to once again become acquainted with its role as an important policymaker." There is certainly justification for this assertion that policy can not consist simply of reliance on "free market rhetoric."

But Nelson then immediately qualified that she was "not calling for increased federal preemption" but rather for dialogue, collaboration, and cooperation. Certainly there is, and always will be, the need for such cooperative efforts in our dual system. But Nelson and NARUC are dead wrong in clinging to the 1914/1934 process. That is not "to set aside narrow parochial interests" but rather to cling to the same insistence on turf priority, however inconsistent it is with the national interests in these changed times.

The only hope for improvement is that NARUC does set aside its narrow parochial interests and call for revision, or that Congress finally does screw up its courage to act. If this does not occur, we will muddle along, hopefully with greater cooperation and sense of responsibility among the regulators, until some scandal or worsening national situation finally compels reform.

"Notwithstanding Section 2(b) . . .": Recent Legislative Initiatives Affecting the Federal–State Balance in Telecommunications Regulation

Jeffrey Tobias
Pike and Fischer, Inc.

As discussed by Noam and Geller in the preceding chapters, the present-day telecommunications marketplace is one singularly ill-suited for a federal–state regulatory dichotomy predicated on geographical boundaries, and one for which greater-than-normal constraints on federal preemptive authority appear especially unwarranted. Yet this is precisely the situation that prevails, with limited exceptions, by virtue of the affirmative reservation of state authority over intrastate services that is codified in Section 2(b) of the Communications Act of 1934, 47 U.S.C. §2(b), as authoritatively construed by the United States Supreme Court in *Louisiana Public Service Commission v. FCC*, 476 U.S. 355 (1986). Section 2(b), the product of a confluence of political circumstances extant in the early decades of this century, is something of an anomaly in the current regulatory landscape. It frequently shackles the FCC from asserting authority to strike down state laws that have the potential to frustrate federal telecommunications policy, although the agency clearly would be entitled to so preempt the conflicting state provisions under well-established Supremacy Clause jurisprudence in the absence of Section 2(b).

In several recent legislative initiatives, however, both Congress and the Executive Branch have signaled a recognition that continued application of Section 2(b) as the governing framework for dividing the spheres of federal and state regulatory authority over new and existing telecommunications technologies may, at least in some instances, disserve the public interest. One major telecommunications law enacted in 1993, and several bills of far-reaching significance introduced and still pending before Congress (with a high probability of passage) as of this writing have explicitly withdrawn areas of telecommunications regulation from the ambit of Section 2(b), often prefacing the sections pertaining to pre-emption with the phrase "Notwithstanding Section 2(b) . . .", or some variant. These provisions, although perhaps not sounding a death knell for Section 2(b), certainly herald a growing Congressional interest in expanding the FCC's preemptive jurisdiction, taking away from the state public service commissions some of what, in the aftermath of *Louisiana PSC*, they could have reasonably believed to be rightfully theirs.

The first major blow to state regulatory prerogatives regarding telecommunications appears in the Omnibus Budget Reconciliation Act of 1993,[1] which was signed into law on August 10, 1993. Among several sections dealing with telecommunications, the Budget Act amended Section 332 of the Communications Act, 47 U.S.C. §332, to mandate regulatory parity for all commercial mobile radio services and to foreclose state regulation of those services in certain areas. A brief synopsis of the key features of Section 332, as amended, is in order here to provide a context for the discussion of Section 332's provisions on preemption that follows.

Commercial mobile radio services (CMRS) are defined in the statute as all mobile radio services that are "provided for profit and make interconnected service available (A) to the public or (B) to such classes of eligible users as to be effectively available to a substantial portion of the public, as specified by regulation by the [Federal Communications] Commission." The term *interconnected service* refers to service that connects the user to the public switched telephone network.

In passing this amendment to Section 332, Congress was concerned that providers of a number of services theretofore classified as private land mobile radio services, most notably wide-area Specialized Mobile Radio (SMR) service and private carrier paging service, had become viable competitors to cellular radio, common carrier paging and other common carrier radio services, but operated under a much more liberalized regulatory scheme. The private land mobile radio service had been created initially to serve the internal communications needs of public safety entities and large businesses, but evolved over the years to include some "private carrier" services that private land mobile licensees were permitted to offer to third parties on a for-profit basis. Especially in the last few years, limitations on end-user eligibility for these private carrier services have been greatly relaxed, further encouraging direct competition with common carrier services. It remained, however, that mobile radio services classified as common carriage were subject to a host of requirements contained in Title II of the Communications Act. Their rates and practices were carefully scrutinized to ensure that they were reasonable and devoid of unjustifiable discrimination, whereas the competing private carriers were free of these Title II requirements. Compounding this disparity in the regulatory treatment of the two types of services was the circumstance that the former version

---

[1]P.L. No. 103-66, §6002, 107 Stat 312, 392.

of Section 332 precluded the states from regulating the rates of private carriers or restricting marketplace entry of private carriers, but did not similarly restrict state regulation of mobile common carrier services.

Radio common carriers attacked the dual regulatory scheme as arbitrary and unfair, claiming that compliance with the Title II requirements and the implementing FCC rules hamstrung them in their efforts to compete vigorously against private carriers. The need to address this issue received added stimulus from the FCC's 1992 proposal to license personal communications services (PCS), a new generation of mobile services that could conceivably be regulated as radio common carriage, private radio, or both, but that, in any event, could provide a highly attractive alternative to many existing mobile service offerings.

Congress' response was to group traditional radio common carriers and competitive private carriers into the new regulatory classification—CMRS—and limit private land mobile regulation to public safety services and other services that are not designed to meet the telecommunications needs of the general public. Under the new Section 332, all CMRS providers are to be regulated as common carriers, but the FCC has been accorded discretion to "forbear" from enforcing certain Title II requirements against CMRS providers if it is satisfied that imposition of those requirements is not necessary. (The agency has in fact exercised this flexibility, holding that a number of Title II provisions will not be applied to CMRS.[2])

Section 332(c)(3) of the Communications Act, as amended by the 1993 Budget Act, specifies that "[n]otwithstanding Section[] 2(b) . . . , no State or local government shall have any authority to regulate the entry of or the rates charged by any commercial mobile service or any private mobile service." (As earlier noted, the ban on state rate or entry regulation of *private* land mobile services antedated passage of the Budget Act.) Congress thus removed from state purview all rate or entry regulation of the fastest growing segment of the telecommunications industry, invalidating, for example, laws governing intrastate cellular rates that several states had enacted, and ensuring federal primacy in PCS policy making. It did so in the belief that the traditional jurisdictional demarcation mandated by Section 2(b) would prove unworkable for mobile services which, in the words of the House Report on the legislation, "by their nature, operate without regard to state lines."[3]

It bears noting that Section 332 does not compel a complete withdrawal of all state regulation of mobile services. First, Section 332 expressly recognizes the right of state authorities to impose nondiscriminatory requirements on CMRS to ensure universal service at affordable rates "where such services are a substitute for land line telephone exchange service for a substantial portion of the communications within such State." Second, any state may petition the FCC for authority to regulate

---

[2]*Implementation of Sections 3(n) and 332 of the Communications Act,* 74 RR 2d 835 (1994).
[3]H.R. 103-11, at 259-260 (1993). The Conference Report is H.R. 103-213.

CMRS rates upon a showing that market conditions in the state will not suffice to ensure that subscribers are protected from unjust and unreasonable rates or unreasonably discriminatory practices. If a state that was already regulating mobile service rates files a petition to continue such regulation, the existing regulatory scheme is permitted to remain in effect until the FCC rules on the petition. But the burden of proof ultimately lies with the state to demonstrate that state rate regulation is necessary. The FCC, in implementing this provision, has held that states will have to "clear substantial hurdles if they seek to continue or initiate rate regulation of CMRS providers."[4] Moreover, any state rate regulation that is authorized on this basis may be ordered discontinued after the passage of a "reasonable period of time" if "any interested party" subsequently files a petition demonstrating that changed circumstances have obviated the need for such regulation. Third, and perhaps most significantly, Section 332 acts as a bar only to rate and entry regulation; by its express terms, it does not preclude state regulation of "other terms and conditions of commercial mobile services." The legislative history[5] indicates that Congress viewed the following as examples of the "terms and conditions" that may still be the subject of state or local regulation: customer billing practices, billing disputes and other consumer protection matters, facilities zoning issues, transfers of control, bundling of service and equipment, and requirements that carriers make capacity available on a wholesale basis for resale. But even with regard to these "terms and conditions" for the offering of commercial mobile services, the states are not guaranteed regulatory primacy; the FCC may still seek to preempt state regulation in any or all of these areas, but it will only be allowed to do so if it can meet the *Louisiana PSC* test for federal preemption under Section 2(b).

In sum, then, the telecommunications preemption provisions of the 1993 Budget Act leave the states with only a modest role in the regulation of mobile services, tilting the balance of federal and state regulation decidedly in favor of the FCC. The balance struck is far different from what would have been the case had Congress simply decided to allow Section 2(b) to supply the governing principles for dividing regulatory authority in this area. Developments occurring in the first session of the 103d Congress suggest that a similar expansion of federal preemptive jurisdiction will occur in the context of the Clinton administration's National Information Infrastructure (NII) initiative and related legislation.

OTHER RECENT CONGRESSIONAL ACTIONS

One of the key telecommunications-related bills to emerge in the 103d Congress is H.R. 3636, the National Communications Competition and Information Infrastructure Act of 1993. Approved by the House Commerce Committee on March 16, 1994, H.R. 3636, if enacted, would overturn the existing cable/telephone

---

[4]*Implementation of Sections 3(n) and 332 of the Communications Act*, 74 RR 2d 835, 843 para. 23 (1994).

[5]H.R. 103-11, at 261 (1993).

company cross-ownership rule that bars LECs from offering video programming directly to subscribers within their service areas. It is also designed to open the "local loop" to competition from cable companies and other entities, imposing new duties on LECs to interconnect with and provide "equal access" to potential competitors.

As initially introduced by Representative Markey, H.R. 3636 has a number of provisions addressing the scope of federal and state regulation. Most significantly, it contained the following:

> STATE PREEMPTION.—Notwithstanding section 2(b), no State or local government may, after one year after the date of enactment of this subsection (A) effectively prohibit any provider of any telecommunication services from providing that or any other such service, or impose any restrictions on entry into the business of providing any such service that is inconsistent with this subsection or any provision of this Act, or any regulation thereunder; (B) prohibit any carrier or other person providing telecommunications services from exercising the access and interconnection rights provided under this subsection; or (C) impose any limitation on the exercise of such rights that is inconsistent with this subsection or the regulation prescribed under this subsection.

This provision would foreclose state regulators from restricting or conditioning new entry into a wide range of wireline telecommunications services, even those for which there may arguably be a segregable intrastate component. In the sweep of its language one can discern an unequivocal rejection of the jurisdictional compromise embodied in Section 2(b). H.R. 3636 has as its vision a new era of largely unfettered competition in the provision of telecommunications services historically provided by de jure or de facto monopolists; in this brave new world, H.R. 3636 seems to say care must be taken to ensure that the federal ambition is not thwarted by state regulators seeking to further narrow interests. The essential prophylactic measure for this purpose is to take Section 2(b) out of play.

H.R. 3636 does acknowledge state concerns. It calls for the establishment of Federal–State joint boards to develop recommendations on a variety of issues, including the preservation of universal service and the jurisdictional cost separations process; provides that nothing in the bill is intended to limit state authority to regulate the allocation of costs for intrastate rate-making purposes; and enjoins the FCC to "coordinate and consult with" state regulatory bodies, among others, in developing network reliability and service quality performance measures that are required by the bill. But these provisions did little to ease state regulators' concerns about the scope of the federal preemption contemplated by the legislation. State officials voiced these concerns at February 1994 hearings on H.R. 3636,[6] and the final version of the bill that passed the Commerce Committee did contain some concessions to the states, principally in the form of specific language

---

[6]*Communications Daily*, February 10, 1994, p. 1.

to clarify state authority to regulate the terms and conditions of the provision of telecommunications and information services.[7] However, the core preemption provisions on entry regulation remained.

Another major telecommunications bill introduced in the 103d Congress is H.R. 3626, the Antitrust Reform Act of 1993, which is designed to phase out the line-of-business restrictions placed on the BOCs by the Modification of Final Judgment. Section 107(c)(2) of the bill, as it was introduced on November 22, 1993, addresses the issue of federal preemption, stating simply, "This Act shall supersede State and local law to the extent that such law would impair or prevent the operation of this Act." This provision appears to be designed to provide a less rigorous test for the assertion of federal preemption than that supplied by Section 2(b). There is, after all, no mention of state prerogatives with regard to intrastate services, and one can assume that the drafters of the legislation would have felt no need to include this language had they intended that preemption of state laws inconsistent with H.R. 3626 simply be in accord with Section 2(b).

It would thus appear from a reading of the provision in isolation that whatever other laws might eventually be superseded under Section 107(c)(2), the one statutory provision that is clearly superseded is Section 2(b). Unfortunately, the matter is confused by the immediately preceding Section 107(c)(1), which asserts cryptically that "[e]xcept as provided in [Section 107(c)(2)], this Act shall not be construed to modify, impair, or supersede *Federal*, State, or local law unless expressly so provided in this Act" (italics added). The upshot is that H.R. 3626's treatment of the preemption issue remains somewhat uncertain, hopefully to be clarified later in the legislative process through redrafting or authoritative statements in the legislative history. But the desire to include in the bill a section specifically dealing with the question of conflicting state regulation does suggest, once again, a dissatisfaction with permitting Section 2(b) to stand as the controlling rule of law. In any event, the H.R. 3626 that is voted on by the full House may be significantly different from the bill originally introduced; in March 1994, separate versions of the bill were voted out by the House Commerce and Judiciary Committees.

CLINTON'S NII AND THE STATES

The Clinton administration has prepared its own legislative package for creation of the National Information Infrastructure. It endorses and builds upon much of what is in H.R. 3636 and H.R. 3626, as well as the counterpart Senate bills. But the administration's proposals in some ways go even further in limiting state regulatory power over telecommunications.

---

[7]*Id.*, March 1, 1994, p. 3.

In its NII White Paper, the administration offered its general view on federal–state relations. It begins cautiously enough:

> Because of the crucial role of the states in protecting ratepayers and addressing economic and technical infrastructure issues in their areas, substantial state jurisdiction over telecommunications must be preserved. However, when national interests are at stake in realizing the benefits of an advanced, interconnected NII, particularly through local competition, national policies, with limited preemptive effect in a few key areas, are necessary.

In elaborating on its proposals, however, the administration indicates that it views clear-cut federal preemption as "necessary" in a number of areas in which the pending House bills have been silent on the question. Thus, for example, the administration embraces H.R. 3636's preemption of state entry regulation of telecommunications and information services providers, but seeks to have added to the bill a provision that would also preempt state and local regulation of "the rates for any service charged by a telecommunications carrier that the FCC finds, or has found, after notice and comment, to lack market power." (The administration would permit states to petition the FCC to retain or regain rate regulation authority under certain circumstances, just as is allowed under Section 332 of the Communications Act.)

Of even greater concern to state regulators is the administration's proposal to add a new Title VII to the Communications Act. Title VII, in essence, would establish a new, streamlined regulatory framework for eligible providers of two-way, broadband digital transmission services—the services viewed as the linchpin of the NII. Although the FCC would be delegated authority to define more precisely which broadband services qualify for streamlined regulatory treatment, the agency would be required to apply its implementing rules without regard to whether the provider of the broadband services is a common carrier or a cable television operator. But to take advantage of this streamlined regulatory treatment, the broadband services provider would have to agree to assume specified obligations pertaining to open access, universal service, and interconnection and interoperability with other service providers.

The administration proposes to preempt state and local regulation of the rates of those Title VII service providers that are determined by the FCC to lack market power, that is, to be subject to competitive pricing constraints. States would still be permitted to regulate the intrastate rates of Title VII service providers that are found to have market power. In addition, according to the NII White Paper, states would be empowered to regulate rates for "other services delivered over the facilities used to furnish Title VII broadband services, in the discretion of the states, subject only to a reserved right of Federal preemption that could be exercised to the extent necessary to avoid conflicts between state regulatory actions and the policies of Title VII."

State regulators are unhappy with Title VII. Citing ambiguities regarding the extent to which Title VII would permit federal preemption of intrastate rate regulation, the Executive Committee of NARUC voted to oppose the initiative on March 3, 1994. Further, although none of the pending telecommunications bills have incorporated the Title VII proposal despite administration lobbying, NARUC's Executive Committee also issued a resolution expressing that body's concern that the pending bills "contain provisions that would alter the jurisdictional authority of [the FCC] and state public utility commissions in regulating interstate and intrastate communications, with some aspects of state regulation preempted and primary responsibility for other activities transferred to the FCC."[8]

Regardless of the ultimate fate of Title VII, it and the other legislative proposals discussed here appear to reflect a growing consensus at the federal level that it is counterproductive to apportion federal and state regulatory authority over telecommunications on the basis of the Section 2(b) interstate/intrastate distinction. Even if Section 2(b) is not repealed and replaced any time soon—and any assessment of the prospects for near-term repeal must take account of the formidable political forces that will be aligned against such action—it seems likely that Section 2(b) will be slowly eviscerated by the continued inclusion in new telecommunications bills of superseding preemption provisions specific to the subject of the legislation. As each new piece of telecommunications legislation with its very own preemption section becomes law, Section 2(b) will continue a downward spiral into irrelevance.

If decisions about the appropriate relationship between federal and state authorities in regulating telecommunications continue to be made by Congress on a case-by-case basis, the expansion of federal preemptive authority will also continue. This may occur in any event, and may very well be in the best interest of the nation, but ad hoc decision making on telecommunications preemption issues will likely lead to considerations of federalism being overwhelmed by the more immediate policy and political imperatives giving birth to the legislation. If the matter at hand is important enough to merit a carefully thought-out Congressional response, federal proponents of the legislation are likely to believe it is also important enough to warrant removing 50 potential obstacles to achievement of the federal goal. Given this circumstance, it may be in the interest of state regulators to now seek from Congress a new statutory provision on the division of federal and state regulatory authority over telecommunications to replace Section 2(b), one that will be sufficiently palatable to federal lawmakers such that its application to new technologies that are the subject of future legislation will be the norm rather than the exception. Any "new" Section 2(b) must define the respective regulatory spheres of the federal government and the states on a more sensible basis than the current, anachronistic interstate/intrastate distinction, but it need not, and should not, be formulated in derogation of legitimate state regulatory interests.

---

[8]As reported in *Communications Daily*, March 4, 1994, p. 2.

8

CONCLUSIONS

Paul Teske
State University of New York
at Stony Brook

Many important U.S. industries are jointly regulated by the national government and by the 50 states. No other regulated industry is more important to the long-run competitiveness of the American economy than telecommunications. State regulators are playing a major role in today's telecommunications industry and no active stakeholder has been able to mount a serious challenge to the ongoing viability of state regulation. Indeed, contributors to this volume have amassed much evidence that states are playing an experimental, laboratory role in a period of substantial technological, economic, and financial uncertainty in this dynamic industry. Even the national government, as exemplified by the FCC's recent interconnection policies, is learning from the states' actions.

Yet, many states are still fighting battles over the essential issues that they faced immediately after divestiture (see Teske, 1990). Local rates are probably still below their costs in many states, leaving more competitive services paying disproportionately high rates. As local competitors gain ground, this subsidy becomes less sustainable, even if the subsidy is smaller than it was a decade ago. This creates incentives for state regulators to drag their feet on competition, explicitly or through regulatory lag.

As states have very gradually changed their pricing and competition policies since the early 1980s, they have also faced a host of new issues as the telecommunications industry becomes more complex and new technological trends create both opportunities and regulatory problems. Some states have approached these new challenges with new ideas and innovations, whereas others have tried to ignore or deny them.

And, these chapters illustrate that serious potential problems with regulatory federalism are looming. As Noam notes, it will be increasingly difficult to categorize telecommunications traffic as intrastate, as diverse networks and new technologies make these boundaries ever more irrelevant (some say they are already irrelevant today, or at least arbitrary). As the telecommunications networks of the future become more complex, modularized, and interconnected, drawing the precise regulatory boundaries will not only become more difficult, it may become a very costly and inefficient exercise.

It would not be unprecedented in America for state regulation to recede or even disappear over time. After being the initiators of regulation in this country, states now do not play an active role in the regulation of railroads, in large part because the 1980 Staggers Act preempted them from doing so. It would not be shocking for this to happen at some point in telecommunications.

The Clinton administration intends to pursue sweeping telecommunications regulatory legislation in its NII initiative. Vice President Gore, the architect of this plan, indicates that part of that legislation, if successful, will give the FCC the power to preempt at least certain kinds of state regulations that limit new companies from offering local telephone or cable television services. On the other hand, Gore has also indicated that he expects to work with the states on other aspects of developing an information age infrastructure.

Because recent court decisions have generally favored the ongoing state role, Congressional action almost certainly is required to reduce the state regulatory role. With the Congressional passage of the 1993 rider to the Budget Reconciliation Act on spectrum allocation, the role of the 19 states that regulated cellular telephone and other wireless service rates, including New York and California, was reduced substantially. Thus, the notion that Congress cannot summon the will to act in the contentious telecommunications arena may be changing.

As Congress does consider its first major telecommunications regulatory action in 60 years, it is also reasonable to ask whether the American reliance on state regulation is at odds with experience in other parts of the world, especially now that Canada has established centralized regulation. For example, although the Western European nations will maintain substantial control over their networks in the coming years, they are coordinating a growing portion of telecommunications activities through the European Community. The direction that Eastern European nations and the newly sovereign Soviet republics will take in the centralization of telecommunications authority is as yet unclear; certainly they have a long way to go simply to catch up in this and other areas. At least for now, the United States appears to be alone in emphasizing a strong role for the states.

There does appear to be the beginning of a consensus at the national level about what telecommunications policy should look like, something that Gormley noted herein as a necessary condition for preemption of the states in favor of a national policy. Most officials seem to support a government policy with competition on a "level playing field," with open architecture and interconnected networks, with prices that reflect economic costs, with antitrust-like regulatory safeguards, and with explicit programs to maintain universal access. As we have seen, however, the implementation and timing of introduction of each of these policy elements will affect stakeholders differently, and are thus controversial and subject to being blocked.

SYNTHESIS OF FINDINGS

We have tried to look both backward and forward for answers to today's difficult jurisdictional questions in telecommunications regulation. In the broader historical picture, Gabel shows how jurisdictional questions are not only academic concerns about the optimal locus of decision making, but critical strategic political choices for the active players. AT&T used state regulation to gain their regulated monopoly, to avoid municipal regulation or ownership, and to maintain a continuing cross-subsidy that was beneficial to the growth of their business and profits. Today it is true that many consumer groups feel that only state regulators are sensitive to their concerns about local residential rate increases, compared to FCC decision makers sitting far away in Washington, DC. Alternatively, even if they do successfully pass an important bill in 1994, members of Congress may wish to avoid making the most difficult pricing and entry decisions by forcing others, including state regulators and perhaps Judge Greene, to continue to do so. Although this may not result in an optimal allocation of policymaking responsibility and accountability, it is not an atypical result in the decentralized American political environment.

Barry Cole finds several models of innovation in the states, although he notes that it is also possible to classify many seemingly different kinds of policies, for example in rate setting, under a few general categories. It is an interesting question about whether state models differ because preferences or tastes for different telecommunications policies truly vary across citizens of the states, or, more likely in my view, because the preferences and decision calculi of political officials vary across states. In any case, states are acting somewhat as a policy laboratory and the variation is real. However, as Megdal notes, these experiments are not being developed or analyzed with a uniform scientific approach. Although some interested stakeholders definitely will be analyzing this cross-state data to argue that certain policies promote the public interest (as well as their own best interest), it is less clear which government agencies have the resources and incentives to do such important analysis. So far, Mueller's (1993) analysis of Nebraska's deregulatory experiment is one of the few careful studies, and it focuses only on a single state over a relatively short period of time.

Teske and Bhattacharya expand on Cole's finding that telecommunications policy is increasingly being influenced by other actors in states beyond the PUC regulators, including legislators, governors, and economic development agencies, as its importance to infrastructure and economic development becomes clearer. The Advisory Commission on Intergovernmental Relations (ACIR, 1990) called this more proactive development "Phase Two" of state telecommunications regulation after divestiture. Teske and Bhattacharya argue that this is a positive development and will tend to increase innovation in the state regulatory environment. New York State's Telecommunications Exchange may be at the leading edge of state efforts to bring a wider array of concerns into state telecommunications policymaking.

The chapters by Megdal and Egan and Wenders present opposing views of the benefits and costs of state regulation, largely using an economic perspective in both cases. Megdal's classic public finance approach thoughtfully considers the benefits of federalism and finds them compelling. She admits that a case can also be made against state regulation but that, on balance, allowing the state variation in policy is the best current choice. Egan and Wenders argue forcefully for rapid telecommunications deregulation generally, to gain the dynamic benefits of competition, and see that as much more likely to occur rapidly under the direction of a single federal agency, rather than the 50 states. They are probably correct about that point (see Teske, 1990).

Haring's comments apply Occam's Razor to the jurisdictional problem in highlighting his own preemption standard that would require a federal burden of proof for preemption that will be difficult in some cases and easier in others, appropriate to the interrelationship of federal and state interests. The other two comments on the pros and cons of state regulation are less willing to consider only traditional economic concerns and instead emphasize values of equity, political participation, and stable institutional choice. Jones notes that economists often ignore institutional issues to which political scientists have been somewhat more attentive. If in fact it is elite decision makers, such as PUC commissioners and staff, legislators, and governors who are responsible for state policy differences rather than differing public preferences across the states, Jones is making an important point. Gormley notes that in the uncertain political and technological environment of telecommunications, and considering equity concerns that state regulators are more likely to be attuned to than federal regulators, state regulation retains considerable advantages for our society.

Noam provides a useful review of the recent legal history of preemption in telecommunications and related industries, then builds upon this to consider how future technological changes will affect the federal–state balance. His general conclusion for state authority is that "the net result is a shrinking share of a shrinking share, which means that state regulation will be under continuous pressure." As a former state regulator, his perspective combines both theoretical and practical political concerns. Noam considers several federal–state possibilities and concludes that cooperation and broad federal policy guidance with room for state experimentation may best serve the national interest.

Geller adds a specific focus on the changing legal analysis of state telecommunications regulation. He argues that although state regulation can be and often is more responsive, localized, and experimental than federal regulation, some policy issues, such as enhanced services, require a single "federal captain" rather than "two hands on the wheel." But Geller is as pessimistic as Noam about the U.S. Congress generating the will to try to provide a better legislative solution to these thorny jurisdictional problems than is provided in the 1934 Communications Act.

Tobias explains why 1994 may be the appropriate time for Congress to act in this important area, with a strong push from the Clinton administration. He suggests that whatever compromise emerges in legislation, the state role is likely to be reduced, at least in some areas, as there is considerable consensus among critical players on this point.

CONCLUSIONS

Two terms that have become somewhat clichéd in public policy debates are currently shaping national goals and approaches to telecommunications. The first is the *information superhighway*. Most media outlets, including newspapers, magazines, and television news, are analyzing some aspect of the information superhighway, whether it is the growth of usage of the Internet or a merger or joint venture of two or more telecommunications, cable television, software, programming, or electronics firms. Despite disagreements about what would really constitute it, technology has been leading in the direction of an information superhighway of some kind, and Vice President Gore's strong advocacy has placed the topic firmly on the public agenda.

What is the state role in building the information superhighway? As Cole noted, most investment decisions by the regulated telephone common carriers are still subject to state regulatory control. The LECs, even when faced with greater competition for their most lucrative customers, are still providing most users with the "access ramps" or "local roads" onto the information superhighway. These are the most expensive parts of the system to build and to maintain, and are likely to remain the least desirable for new competitors because of their ubiquity and expense.

Ironically, at the same time the information superhighway is being discussed, new wireless technologies are forcing analysts to question the assumption that more traditional wireline companies will dominate the future expansion of telecommunications markets. No one knows exactly how the future of telecommunications technology will evolve, except to be sure that it will be different from the past and present industry. The 1993 Congressional decision to minimize the state role in setting cellular and mobile telephone services rates is a telling signal about how national policies are likely to be shaped with regard to preemption of this wireless technology.

The second critical term that has become clichéd, but that is nevertheless important, is *globalism*. International trade increases every year and more and more of it is linked, and even created, by sophisticated global telecommunications systems. For the sake of U.S. economic development, key decision makers want U.S. firms to be well positioned in this competitive environment. Distributed computing, satellites, the deployment of fiber optics, the Internet, wireless communications, and the merging of various forms of digital technology create a technological environment that will profoundly shape business and social interactions across the world. Mega-mergers among telecommunications carriers, cable TV firms, television networks, and electronics firms are shaping the environment for a global future where analysts believe that only firms able to exploit large economies of scale can succeed. The largest telecommunications providers and the largest users are thinking in terms of international, rather than just national, networks.

Because each country has different telecommunications regulatory policies, differences that certainly far exceed the differences in state regulation in the United States (see Duch, 1991), telecommunications providers and users face a challenging task. But, many of the same competitive pressures on AT&T in the 1970s in the United States are being felt in other countries and in the international environment. Inevitably, the international monopoly of Intelsat is facing major competition from other satellite firms and from new technologies. With such a focus on international firms and international policies, is it inappropriate for some of the most important U.S. firms to be partially bound by state regulation?

Within the United States, interstate commerce is increasingly important in many industries. Old assumptions about state regulation are being questioned in such diverse industries as insurance and banking. Truly intrastate commerce is much more rare and likely to become very rare in telecommunications.

Until we get to that point, however, there is still a substantial amount of intrastate telecommunications traffic, and there are still some political advantages to regulating at that level, which suggest that it will not be easy to remove, even if it becomes desirable to do so.

An important conclusion is that state regulation has served a valuable purpose but will likely be reduced in its role as the breathtaking technological changes continue to occur. State regulation will not end before the turn of the century or perhaps for years, or even decades, beyond that. But it almost certainly will lose much of its role over time. How that transition is completed, how states cooperate with national objectives and with one another, and how states handle the important policy choices they still control, will greatly affect the success of the U.S. telecommunications industry and, indeed, of the U.S. economy in global competition.

In the meantime, Congress should use the emerging consensus for competition and the protection of captive consumers to develop a broad telecommunications policy that sets up clearer preemption standards for the FCC. Congress should explicitly preempt or give the FCC the power to preempt truly national issues, as identified by Haring, Geller, and others as information services, issues involving wireless spectrum usage, and other areas. State policy laboratories should be encouraged in other areas of pricing and regulation, as long as they do not adversely impact interstate commerce, and the administration should work with the states to develop various options for the information superhighway. As Noam notes, "there is much work still to be done." Because the states are clearly going to continue to be involved for some time, cooperative policy will work better than jurisidictional battling. The establishment of national regulation and pre-emption of state authority based on pragmatic policy goals rather than narrow legalistic reasons would be a step in the proper direction.

References

Advisory Commission on Intergovernmental Relations. (1990). *Intergovernmental regulation of Telecommunications*. Washington, DC: Author.

Arnheim, L. (1988). *Telecommunications infrastructure and economic development in the Northeast-Midwest Region*. Washington, DC: Northeast-Midwest Institute.

Baird, E. G. (1934). *Telephone rate making: Judicial restrictions upon commission distribution of the telephone price burden*. Blanchester, OH: Brown.

Bator, F. (1971). The anatomy of market failure. In W. Breit & H. Hochman (Eds.), *Readings in microeconomics* (pp. 518–537). NY: Norton.

Baumol, W., Panzar, J., & Willig, R. (1982). *Contestable markets and the theory of industry structure*. San Diego: Harcourt Brace Jovanovich.

Berry, J. (1989). *The interest group society*. Glenview, IL: Scott, Foresman.

Bernstein, M. (1955). *Regulating business by independent commission*. Princeton, NJ: Princeton University Press.

Blau, R. (1990). *Judicial policymaking in the U.S. telecommunications industry and its implications*. Unpublished manuscript, Bell South Corporation, Atlanta, GA.

Blazar, W. (1985). *Infrastructure support for a new generation of economic activity: Telecommunications*. Chicago: American Planning Association.

Bowman, A., & Kearny, R. (1986). *The resurgence of the states*. Englewood Cliffs, NJ: Prentice-Hall.

Brock, G. (1981). *The telecommunications industry: The dynamics of market structure*. Cambridge, MA: Harvard University Press.

Coalition of Northeastern Governors. (1987, April). *Telecommunications and economic development: A regional view*. (Transcript of Coalition of Northeastern Governors Policy Research Center)

Cohen, J. (1992). *The politics of telecommunications regulation: The state and the divestiture of AT&T*. Armonk, NY: M.E. Sharpe.

Coll, S. (1986). *The deal of the century: The breakup of AT&T*. New York: Atheneum.

Coopers and Lybrand, Inc. (1989, February). *The impact of the emerging intelligent network in New York State*. New York, NY: Author.

Crandall, R. (1991). *After the breakup: U.S. telecommunications in a more competitive era.* Washington, DC: Brookings Institution.

Derthick, M., & Quirk, P. (1985). *The politics of deregulation.* Washington, DC: Brookings Institution.

Duch, R. (1991). *Privatizing the economy: Comparative telecommunications regulation.* Ann Arbor: University of Michigan Press.

Edelman, M. (1964). *The symbolic uses of politics.* Champaign: University of Illinois Press.

Eisinger, P. (1988). *The rise of the entrepreneurial state.* Madison: University of Wisconsin Press.

Elazar, D. (1974). The new federalism: Can the states be trusted? *The Public Interest, 35,* 89–102.

Entman, R. (1988). State telecommunications regulation: Developing consensus and illuminating conflicts. Wye, MD: *Aspen Institute Program in Communication and Society Report.*

Erickson, H. (1915). The advantage of state regulation. *Annals of the American Academy, 57,* 123–162.

Fainsod, M., & Gordon, L. (1941). *Government and the American economy.* New York: Norton.

Federal Communications Commission. (1939). *Investigation of the telephone industry in the United States.* Washington, DC: U.S. Government Printing Office.

Federal Communications Commission. (1990). *Update on quality of service for the Bell Operating companies.* Washington, DC: U.S. Government Printing Office.

Federal Communications Commission. (1991). *Semiannual report on telephone trends.* Washington, DC: U.S. Government Printing Office.

Ferejohn, J., & Shipan, C. (1989). Congress and telecommunications policymaking. In P. Newberg (Ed.), *New directions in telecommunications policy* (pp. 301–314). Durham, NC: Duke University Press.

Fosler, S. (1988). *The new economic role of the American states.* New York: Oxford University Press.

Gabel, D. (1987). *The evolution of a market: The emergence of regulation in the telephone industry of Wisconsin, 1893–1917.* Doctoral dissertation, University of Wisconsin-Madison.

Gabel, D. (1990). Divestiture, spin-offs, and technological change in the telecommunications industry—A property rights analysis. *Harvard Journal of Law and Technology, 3,* 75–102.

Gabel, D. (1994). Competition in a network industry: The telephone industry, 1894–1910. *Journal of Economic History, 54,* 535–561.

Garber, S. (1990). *Joint federal and state regulation with conflicting regulatory objectives: The case of telecommunications network access charges.* Working paper, Carnegie-Mellon University, Pittsburgh, PA.

Garber, S., & Peterson, T. L. (1990). *Pricing decisions of state telecommunications regulators under uncertainty.* Working paper, Carnegie-Mellon University, Pittsburgh, PA.

Garnet, R. (1985). *The telephone enterprise: The evolution of the Bell System's horizontal structure, 1876–1909.* Baltimore: Johns Hopkins University Press.

Geller, H. (1989). Reforming the federal telecommunications policy process. In P. Newberg (Ed.) *New directions in telecommunications policy* (pp. 315–333). Durham, NC: Duke University Press.

Gormley, W. (1983). *The politics of public utility regulation.* Pittsburgh: University of Pittsburgh Press.

Gormley, W. (1986). Issue networks in regulatory federalism. *Polity, 18,* 595–620.

Gormley, W. (1987). Intergovernmental conflict on environmental policy: The attitudinal connection. *Western Political Quarterly, 40,* 285–303.

Gormley, W. (1989). *Taming the bureaucracy: Muscles, prayers and other strategies.* Princeton, NJ: Princeton University Press.

Griffin, J. (1982). The welfare implications of externalities and price elasticities for telecommunications pricing. *Review of Economics and Statistics, 64,* 59–66.

Hanneman, G. (1986, March). Applying the idea: Telecommunications and economic development. *NATOA News,* pp. 1–10.

Hardy, A. (1980, December). The role of the telephone in economic development. *Telecommunications Policy, 4,* 278–286.

Haring, J., & Levitz, K. (1989). The law and economics of federalism in telecommunications. *Federal Communications Law Journal, 41*, 261–330.

Harris, R. (1988a). *California telecommunications policy for the twenty-first century: A report to the Calfornia economic development corporation.* Berkeley: University of California.

Harris, R. (1988b, July). *Telecommunications policy in Japan: Lessons for the U.S.* Paper presented at Rutgers University Conference on Public Utility Economics and Regulation, Monterey, CA.

Hudson, H. (1990, October). *Telecommunications policy: The state role. A national overview.* Paper presented to the Eighteenth Annual Telecommunications Policy Conference, Airlie, VA.

Hughes, J. (1977). *The government habit: Economic controls from colonial times to the present.* New York: Basic Books.

Huntington, S. (1952). The marasmus of the ICC: The commission, the railroads, and the public interest. *Yale Law Journal, 61*, 467–509.

Illinois Commerce Commission. (1992). *Telecommunications free trade zones: Crafting a model for local exchange competition* (Report). Chicago: Author.

Jacobson, R. (1989). *An "open" approach to information policy making: A case study of the Moore universal Telephone Service Act.* Norwood, NJ: Ablex.

Jaffe, L. (1954). The effective limits of the administrative process: A reevaluation. *Harvard Law Review, 67*, 1105–1135.

Kahn, A. (1970). *The economics of regulation.* New York: Wiley.

Kahn, A. (1990, September 13). Telecommunications, competitiveness and economic development— What makes us competitive? *Public Utilities Fortnightly*, pp. 16–23.

Landau, M. (1969). Redundancy, rationality, and the problem of duplication and overlap. *Public Administration Review, 29*, 346–358.

Larson, A. C., Makarewicz, T. J., & Monson, C. S. (1989). The effect of subscriber line charges on residential telephone bills. *Telecommunications Policy, 13*, 337–354.

Leach, R. (1970). *American federalism.* New York: Norton.

Lindblom, C. (1977). *Politics and markets.* New York: Basic Books.

Markey, E. J. (1990, April 23). *Telecommunications Week*, p. 1.

McCraw, T. (1984). *Prophets of regulation.* Cambridge, MA: Harvard University Press.

McCubbins, M. (1985). Legislative design of regulatory structure. *American Journal of Political Science, 29*, 721–749.

Megdal, S. B., & Lain, D. (1988). A comparison of alternative methods for regulating local exchange companies. In *Proceedings of the sixth NARUC biennial regulatory information conference* (Vol. 3, pp. 11–41). Washington, DC: NARUC.

Missouri Public Utility Commission. (1991). *Network modernization and incentive regulation.* Jefferson City, MO: Author.

Moe, T. (1989). The politics of bureaucratic structure. In. J. Chubb & P. Peterson (Eds.), *Can the government govern?* (pp. 267–329). Washington, DC: Brookings Institution.

Moss, M. (1986). A new agenda for telecommunications policy. *New York Affairs, 9*(3), 86.

Moss, M. (1987). Telecommunications and the economic development of cities. In W. Dutton, J. Blumler, & K. Kraemer (Eds.), *Wired cities—Shaping the future of communications* (pp. 139–153). Boston: G.K. Hall.

Mount-Campbell, C. A., & Choueiki, H. M. (1987). *A method to estimate long-run marginal cost of switching for basic telephone service customers.* Columbus, OH: National Regulatory Research Institute.

Mueller, D. (1991). *Public choice.* New York: Basic Books.

Mueller, M. (1993). *Telephone companies in paradise: A case study in telecommunications deregulation.* New Brunswick, NJ: Transaction.

National Telecommunications and Information Administration. (1988). *Telecom 2000.* Washington, DC: U.S. Department of Commerce.

New York City Partnership. (1990). *The $1 trillion gamble: Telecommunications and New York's economic future.* New York: Author.

Niskanen, W. (1971). *Bureaucracy and representative government.* Chicago: Aldine.

Noll, R. (1986). *Managing the transition to competition in telecommunications.* Working paper, Stanford University, Palo Alto, CA.

Noll, R. (1989). *Additional comments on statement of goals and strategies for state telecommunications regulation.* Aspen Institute.

Noll, R., & Owen, B. (1989). United States v. AT&T: An interim assessment. In S. Bradley & J. Hausman (Eds.), *Future competition in telecommunications* (pp. 57–89). Boston: Harvard Business School Press.

Oates, W. (1972). *Fiscal federalism.* New York: Harcourt Brace Jovanovich.

Olson, M. (1982). *The rise and decline of nations: Economic growth, stagnation and social rigidities.* New Haven, CT: Yale University Press.

Osborne, D. (1988). *Laboratories of democracy.* Cambridge, MA: Harvard Business School Press.

Owen, B., & Braeutigam, R. (1980). *The regulation game: Strategic use of the administrative process.* Cambridge, MA: Ballinger.

Parker, E., Hudson, H., Dillman, D., & Roscoe, A. (1989). *Rural America in the information age: Telecommunications policy for rural development.* Lanham, MD: Aspen Institute and University Press of America.

Peltzman, S. (1976). Toward a more general theory of regulation. *Journal of Law and Economics, 19*, 211–240.

Peterson, P., & Rom, M. (1990). *Welfare magnets.* Washington, DC: Brookings Institution.

Platt, H. C. (1989). The cost of energy: Technological change, rate structures, and public policy in Chicago, 1880–1920. *Urban Studies, 26*, 32–45.

Pollard, W. (1990). *An examination of the application of peak methods to allocate a revenue requirement for intrastate telephone services.* Columbus, OH: National Regulatory Research Institute.

Posner, R. (1974). Theories of economic regulation. *Bell Journal of Economics and Management Science, 5*, 335–358.

Posner, R. (1982). Toward an economic theory of federal jurisdiction. *Harvard Journal of Law and Public Policy, 6*, 40–57.

Rohlfs, J. (1979). *Economically efficient Bell System prices* (Economic Discussion Paper #138). Morristown, NJ: Bell Laboratories.

Saunders, R., Warford, J., & Wellenius, B. (1983). *Telecommunications and economic development.* Baltimore: Johns Hopkins University Press.

Schattschneider, E. E. (1960). *The semisovereign people: A realist's view of democracy in America.* New York: Holt, Rinehart & Winston.

Schmandt, J., Williams, F., & Wilson, R. (1989). *Telecommunications policy and economic development: The new state role.* New York: Praeger.

Scholz, J. (1981). State regulatory reform. *Policy Studies Review, 1*, 347–360.

Smilor, R., Kozmetsky, G., & Gibson, D. (Eds.). (1988). *Creating the technopolis: Linking technology commercialization and economic development.* Cambridge, MA: Ballinger.

Stehman, W. (1925). *The financial history of the American Telephone and Telegraph Company.* Boston: Houghton-Mifflin.

Stigler, G. (1971). The theory of economic regulation. *Bell Journal of Economics and Management Science, 2*, 3–21.

Stigler, G. (1975). *The citizen and the state: Essays on regulation.* Chicago: University of Chicago Press.

Stone, A. (1989). *Wrong number: The breakup of AT&T.* New York: Basic Books.

Teske, P. (1987). *State telecommunications regulation: Assessing issues and options in the midst of changing circumstances.* Wye, MD: Aspen Institute Program on Communications and Society.

Teske, P. (1990). *After divestiture: The political economy of state telecommunications regulation.* Albany: State University of New York Press.

Teske, P. (1991a, February). Interests and institutions in state regulation. *American Journal of Political Science, 35*, 139–154.

Teske, P. (1991b, January). Rent-seeking in the deregulatory environment: State telecommunications. *Public Choice, 68*, 235–243.

Teske, P. & Gebosky, J. (1991, October). Local telecommunications competitors: Strategy and policy. *Telecommunications Policy, 15*, 429–436.

Teske, P., Best, S., & Mintrom, M. (1994). The economic theory of regulation and trucking deregulation. *Public Choice, 79*, 247–256.

Theodore Barry & Associates. (1991). *Evaluation of Alabama's Rate Stabilization and Equalization Plan*. New York: Author.

Thompson, W. C., & Smith, W. R. (1941). *Public utility economics*. New York: McGraw-Hill.

U.S. Congressional Office of Technology Assessment. (1990). *Critical connections: Communication for the future* (OTA-CIT-407). Washington, DC: U.S. Government Printing Office.

Weinstein, J. (1968). *The corporate ideal in the liberal state: 1900–1918*. Boston: Beacon Press.

Wenders, J. (1987). *The economics of telecommunications*. Cambridge, MA: Ballinger.

Wenders, J., & Egan, B. (1986). The implications of economic efficiency for US telecommunications policy. *Telecommunications Policy, 10*, 33–40.

Wheat, C. I. (1938). The regulation of interstate telephone rates. *Harvard Law Review, 51*, 846–883.

Williams, F., & Barnaby, J. (1992). *Telecommunications regulation and economic development: A view of the states*. (Report of the Center for Research on Communication Technology and Society) Austin: University of Texas.

Wilson, R., & Teske, P. (1990, May). Telecommunications and economic development: The state and local role. *Economic Development Quarterly, 4*, 158–174.

Author Index

Subject Index

159

ABOUT THE AUTHORS

Mallika Bhattacharya holds PhD and MA degrees in Political Science from the State University of New York at Stony Brook. Her dissertation analyzes the motivations for legislative delegation in state telecommunications and trucking regulation.

Barry Cole is currently on leave from the Columbia University Institute for Tele-Information, as Director of the Michigan Ensemble Theater in Traverse City, Michigan, and an Adjunct Professor of Telecommunications at Michigan State University. He holds PhD and MA degrees in Mass Communications from Northwestern University and a BA from the University of Pennsylvania. Cole was previously Director of Programs for the Columbia Institute for Tele-Information, Professor of Telecommunications at the University of Pennsylvania and Indiana University, and a consultant to several federal communications organizations. He is editor of *After the Breakup: Assessing the New Post-AT&T Divestiture Era* (Columbia University Press, 1991) and was co-recipient of two national book-of-the-year awards for *Reluctant Regulators: The FCC and the Broadcast Audience* and *Federal Funding of Children's Television Programming*.

Bruce Egan is a special consultant and Affiliated Research Fellow at the Columbia Institute for Tele-Information. He holds an MA degree in Economics and a BA from Southern Illinois University. He previously worked as an economist for Southwestern Bell and Bell Communications Research. He has consulted for the U.S. Congress, the EEC, the OECD, the United Nations, and other organizations. He is the author of several articles on fiber optic deployment and the 1991 book *Information Superhighways: The Economics of Advanced Public Communications Networks* (Artech Publishers).

David Gabel is Associate Professor of Economics at Queens College. He holds a PhD in Economics from the University of Wisconsin. His current research is in the areas of the economics of regulation, cost allocation procedures, and pricing of telecommunications services, as well as evaluating the origin of and continued need to regulate the industry. He has worked for the Massachusetts Department of Public Utilities, the Wisconsin Public Service Commission, and AT&T. He has published in several economics and telecommunications journals.

165

Henry Geller is Communications Fellow with the Markle Foundation and teaches at both George Washington University and Duke University. He previously headed the Washington Center for Public Policy Research, was Assistant Secretary for Communication and Information Administration in the Department of Commerce, and was Special Assistant to the Chair and General Counsel of the FCC.

William Gormley is Professor of Political Science and Public Policy at Georgetown University. He holds a PhD from the University of North Carolina at Chapel Hill. He is the author of several articles and books on regulation, public policy, federalism, and bureaucracy, including the 1983 *The Politics of Public Utility Regulation* (University of Pittsburgh Press). He previously taught at the University of Wisconsin. His current research analyzes the effectiveness of state child care regulations.

John Haring is a principal in Strategic Policy, Inc. He holds a PhD in Economics from Yale University. Haring previously served as Chief Economist for the Federal Communications Commission and head of the Commission's Office of Plans and Policy. He previously performed economic and regulatory research for the Federal Trade Commission, the Civil Aeronautics Board, and the U.S. Department of Justice, and taught on the Economics faculties of the Universities of Virginia and Maryland.

Douglas N. Jones is Director of the National Regulatory Research Institute and Professor of Regulatory Economics at Ohio State University. He holds PhD and MA degrees from Ohio State and a BA from the University of New Hampshire. He served in several advisory positions in Washington, D.C., including the Department of Commerce and Congressional Research Service. His publications on utility regulation have appeared in professional and trade journals, proceedings, and congressional committee reports.

Sharon B. Megdal is President and founder of MegEcon Consulting Group, specializing in economic policy consulting. She holds PhD and MA degrees in Economics from Princeton University and a BA from Douglass College. She previously served as a commissioner on the Arizona Corporation Commission, as a Director of Tucson Electric Power Company, on the Resolution Trust Corporation Council of Arizona, and on the Economics faculty at the University of Arizona. Much of her recent research focuses on the economics and policy issues related to state telecommunications regulation.

Eli Noam is Director of the Columbia Institute for Tele-Information and Professor at the Columbia Graduate School of Business. He holds a PhD in Economics and a JD law degree and a BA from Harvard University. He has advised numerous government organizations on telecommunications and information strategies. Noam is the author of more than 100 publications, including several books on telecommunications issues. Recent books include *Telecommunications in Europe, Television in Europe, Telecommunications in the Pacific,* and *International Trade in Film and Television.* He is editor of the forthcoming book *Private Networks, Public Objectives.*

Paul Teske is Associate Professor of Political Science at SUNY Stony Brook. He holds a PhD and MPA degrees in Public Affairs from Princeton University's Woodrow Wilson School and a BA from the University of North Carolina at Chapel Hill. Teske has analyzed telecommunications policy for the N.Y. City Department of Telecommunications, the N.Y. State Telephone Association, Teleport Communications Group, and the U.S. Congressional Office of Technology Assessment. He is the author of several journal articles on telecommunications, wrote the 1990 book *After Divestiture: The Political Economy of State Regulation* (SUNY Press), and is co-author of the 1995 book *Public Entrepreneurs: Agents for Change in American Government* (Princeton University Press).

Jeffrey Tobias is the managing editor of Pike & Fischer's *Radio Regulation*, the leading resource for research on federal communications law. He is also managing editor of a number of other Pike & Fischer publications and electronic services covering various areas of FCC regulation.

He holds a BA from the University of Rochester and a JD from the George Washington University National Law Center.

John Wenders is Professor of Economics at the University of Idaho. He holds PhD and MA degrees from Northwestern University and a BA from Amherst College. He is the author of the 1987 book *The Economics of Telecommunications*. He has considerable experience providing economic consulting advice to state public utility regulators.

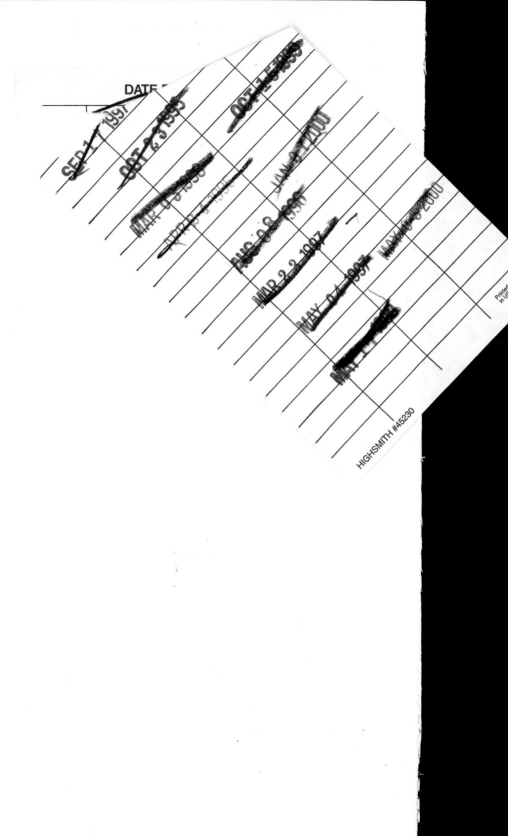
DATE

SEP 17 1997

OCT 23 1998

OCT 12 1999

JAN 26 2000

MAR 08 1999

APR 17 1998

AUG 08 1990

JAN 25 2000

MAY 02 2000

MAR 2 3 1997

MAY 07 1997

MAY 15

HIGHSMITH #45230

Printed
in USA